Visual Perception and Control of Underwater Robots

Visual Perception and Control of Underwater Robots

Junzhi Yu, Xingyu Chen, and Shihan Kong

CRC Press
Taylor & Francis Group
Boca Raton London New York

CRC Press is an imprint of the
Taylor & Francis Group, an **informa** business

First edition published 2021
by CRC Press
6000 Broken Sound Parkway NW, Suite 300, Boca Raton, FL 33487-2742

and by CRC Press
2 Park Square, Milton Park, Abingdon, Oxon, OX14 4RN

Library of Congress Cataloging-in-Publication Data

Names: Yu, Junzhi (Writer on robotics), author. | Chen, Xingyu (Writer on robotics), author. | Kong, Shihan, author.
Title: Visual perception and control of underwater robots / Junzhi Yu, Xingyu Chen, Shihan Kong.
Description: First edition. | Boca Raton, FL : CRC Press/Taylor & Francis Group, LLC, 2021. | Includes bibliographical references. | Summary: "This book covers theories and applications from aquatic visual perception and underwater robotics. Within the framework of visual perception for underwater operations, image restoration, binocular measurement, and object detection are addressed. More specifically, the book includes adversarial critic learning for visual restoration, NSGA-II-based calibration for binocular measurement, prior knowledge refinement for object detection, analysis of temporal detection performance, as well as the effect of the aquatic data domain on object detection. With the aid of visual perception technologies, two up-to-date underwater robot systems are demonstrated. The first system focuses on underwater robotic operation for the task of object collection in sea. The other one is an untethered biomimetic robotic fish with a camera stabilizer, its control methods based on visual tracking. The authors provide a self-contained and comprehensive guide to understand underwater visual perception and control. Bridging the gap between theory and practice in underwater vision, the book features implementable algorithms, numerical examples, and tests, where codes are publicly available. Meanwhile, the mainstream technologies that are covered in the book include deep learning, adversarial learning, evolutionary computation, robust control, and underwater bionics. Researchers, senior undergraduate and graduate students, and engineers dealing with underwater visual perception and control will benefit from the book"-- Provided by publisher.
Identifiers: LCCN 2020039839 (print) | LCCN 2020039840 (ebook) | ISBN 9780367695781 (hardcover) | ISBN 9781003144281 (ebook)
Subjects: LCSH: Remote submersibles. | Robots--Control systems. | Robot vision.
Classification: LCC TC1662 .Y834 2021 (print) | LCC TC1662 (ebook) | DDC 623.82/05--dc23
LC record available at https://lccn.loc.gov/2020039839
LC ebook record available at https://lccn.loc.gov/2020039840

ISBN: 978-0-367-69578-1 (hbk)
ISBN: 978-0-367-70030-0 (pbk)
ISBN: 978-1-003-14428-1 (ebk)

Typeset Minion
by Deanta Global Publishing Services, Chennai, India

Contents

CHAPTER 1

Introduction

1.1 RESEARCH BACKGROUND

The ocean contains abundant natural resources, and also provides a broad space for human development. Constantly deepening the understanding of the ocean and constantly tapping the potential of the ocean is the only way for a country and society to develop. In the era of ocean exploration, as an important tool for humans to go to the ocean, underwater robots emerged and developed gradually. For example, Yuan et al. developed an underwater glide robot that can perform deep-sea exploration tasks [1]; Cai et al. built a hybrid-driven underwater robot that can achieve high maneuverability in a marine environment [2]; Gong et al. designed underwater software whose manipulator has been successfully applied to the target capture task of marine animals [3]. Relying on advanced mechanical structures, control strategies, and perception methods, underwater robots have gradually replaced humans in their motion capabilities, can perform some complex underwater operations, and have played an important role in ocean-oriented resource development and ecological protection.

For instance, "marine ranch" is an emerging economic model that predicts the use of the marine environment to cultivate and manage fishery resources. At present, observation and collection of undersea products primarily rely on artificial underwater operations, which seriously damage the health of the workers (see Figure 1.1). In addition, seafood can also be collected by trawl fishing, but this manner causes a long period of damage to the soil of the seabed, violating the concept of ecological environment to some extent. For observation and collection tasks for marine life, robots

FIGURE 1.1   Marine ranch located at Zhangzidao, Dalian, China.

need two capabilities. On the one hand, robots should be equipped with a visual perception system, capable of optically identifying and positioning biological targets on the seabed. On the other hand, robots should be equipped with a soft mechanical arm, which can carry out non-destructive grasping of biological targets with small Young's modulus. Replacing traditional artificial underwater activities with single robotic operations will effectively alleviate the problems of manpower shortage and operational risk. Replacing traditional trawl fishing with clustered robotic operations can effectively reduce the damage to marine ecology. It can be seen that the underwater robot requires a high level of both motion performance and perception ability.

In recent years, under the guidance of computer, internet, and multimedia technology, artificial intelligence technology has developed rapidly, and it has been widely used in many real-world scenarios such as intelligent video surveillance, smart home, and smart retail. In particular, in the field of computer vision, artificial intelligence technology has effectively solved the problems of image classification, object detection, and object segmentation [4]. Among them, object detection technology can provide the robot with the information of "what" and "where" the object is. In this sense, object detection task can be divided into two sub-tasks, namely classification and localization. Among them, classification task is responsible for judging the possibility of the appearance of the object of interest, and localization task usually outputs a rectangular bounding box to represent the position of the object in the image or video. With the advent of deep convolutional neural networks (CNN) [5], the performance of object detection has been significantly improved. At the same time, the establishment of datasets (e.g., PASCAL VOC [6], MS COCO

[7], ImageNet [8]) also significantly promote the development of the target detection field. The two-stage detector represented by Faster RCNN [9] divides object detection into two stages of region proposal and detection. This mode achieves a high detection accuracy. The single-stage detectors represented by YOLO [10] and SSD [11] abandon the region proposal stage and use end-to-end convolutional neural networks to directly classify and locate objects based on prior knowledge. In contrast, this detection mode has a faster inference speed.

In autonomous robot systems, scene perception servers as the basis for robot decision-making and control, and object detection is the basis of scene perception. As illustrated in Figure 1.2, in the robot task for marine pastures, target detection is the basic sensing task. On the basis of target detection, upper-level perception tasks such as multi-target tracking, key point extraction, and 3D measurement can be carried out. Eventually, the perception information will guide the robot's decision-making and control, so as to achieve autonomous operation. Although object detection methods have been extensively and deeply researched, there are few works to study target detection in robot tasks. On the one hand, unlike images, the robot's visual information has strong temporal consistency, and traditional object detection methods are difficult to capture this temporal information. Therefore, how to use the temporal information to improve the perception level is the key problem of object detection in robot tasks. On the other hand, unlike the dataset, the real-world visual signal often suffers from noise interference, and the visual degradation in the underwater environment is particularly serious. Therefore, the performance of object detection in low-quality real-scene data needs further exploration.

From the perspective of underwater robotics, this chapter summarizes six aspects of research status of underwater visual restoration, static object detection, temporal object detection, commonly used datasets, metrics of object detection, and underwater stereo measurement.

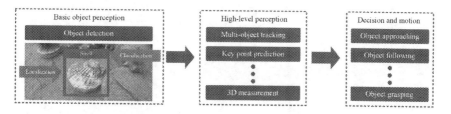

FIGURE 1.2   Object detection in autonomous system.

1.2 REVIEW OF UNDERWATER VISUAL RESTORATION

1.2.1 Formation of Underwater Image

As shown in Figure 1.3, Schechner and Karpel conducted a mechanistic study on the formation of underwater vision [12] and modeled the propagation process of underwater optical signals, which includes direct propagation and forward scattering. In the process of direct propagation, part of the energy is absorbed by water, forming a color distortion and low contrast vision. The optical signal after direct propagation can be expressed as an exponential decay process.

$$d = e^{-\eta z} l, \tag{1.1}$$

where η is attenuation coefficient; z is object-to-camera distance; l is the original optical signal.

Forward scattering causes haziness, and scattered signal f can be formulated as

$$f = d * g_z, \tag{1.2}$$

where $*$ is convolutional operation and g_z is point spread function.

On the other hand, Schechner and Karpel introduced backscattered light b in the visual formation model [12]. The surrounding light source enters the line of sight during the backward scattering process. Therefore, visual degradation is the result of an integrated process, and the degraded visual signal q can be expressed as

$$q = d + f + b. \tag{1.3}$$

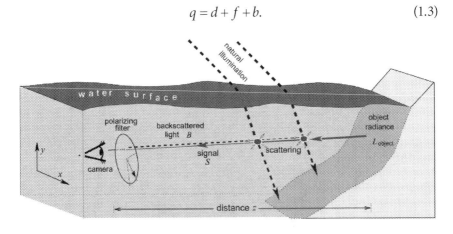

FIGURE 1.3   Object detection in autonomous system [12].

1.2.2 Visual Restoration Based on Image Formation Model

As illustrated in Figure 1.4, some existing methods for restoring underwater images are based on an image formation model (IFM) [13–16], where the background light and transmission map should be estimated in advance. IFM can be formulated as

$$q_\lambda(p) = l_\lambda(p)t_\lambda(p) + BL_\lambda(1 - t_\lambda(p)). \tag{1.4}$$

where p is pixel position and $\lambda \in \{R, G, B\}$ denotes wave length; q_λ is degenerated optical signal; l_λ is original optical signal; BL_λ denotes background light (BL); t_λ is transmission map (TM), denoting the ratio of optical decay. t_λ is usually formulated as an exponential decay, i.e., $t_\lambda = e^{-\eta\lambda} di$, where η is attenuation coefficient; di is object-to-camera distance. It is seen that BL_λ and t_λ can be hard to measure, so information estimation is required for them. For example, dark channel prior (DCP) [17] is one of popular estimation pipelines, which can be given as

$$q_{dark}(p) = \min_{i \in \Omega(p)} \left\{ \min_{\lambda \in \{R,G,B\}} q_\lambda(i) \right\}. \tag{1.5}$$

where $\Omega(p)$ denotes pixel neighborhood centered by p.

According to [14], BL can be determined by the top 0.1% brightest pixels in q_{dark}, namely $p_{0.1\%}$. Thus, BL can be formulated as

$$\widehat{BL_\lambda} = q_\lambda \left(\mathrm{argmax}_{p \in p_{0.1\%}} \sum_{\lambda \in \{R,G,B\}} q_\lambda(p) \right). \tag{1.6}$$

In addition, according to [14], q_{dark} of haze-free images usually equals zero. Assuming $t_R = t_G = t_B = t$, the following equation can be derived by (1,1):

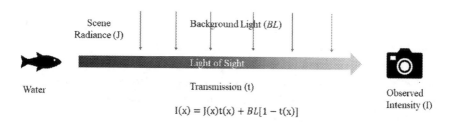

FIGURE 1.4   IFM model [15].

$$\min_{i\in\Omega(p)} \min_{\lambda\in\{R,G,B\}} \frac{q_\lambda(i)}{\widehat{BL_\lambda}} = \min_{i\in\Omega(p)} \min_{\lambda\{R,G,B\}} \frac{l_\lambda(i)}{\widehat{BL_\lambda}} t_\lambda(i) + 1 - t(p). \quad (1.7)$$

where $\hat{t}(p) = \min_{i\in\Omega(p)} \min_{\lambda\in\{R,G,B\}} t_\lambda(i)$. Because $\min_{i\in\Omega(p)} \min_{\lambda\{R,G,B\}} \frac{l_\lambda(i)}{\widehat{BL_\lambda}}$ $= 0$, $\hat{t}(p)$ can be formulated as

$$\hat{t}(p) = 1 - \min_{i\in\Omega(p)} \min_{\lambda\in\{R,G,B\}} \frac{q_\lambda(i)}{\widehat{BL_\lambda}}. \quad (1.8)$$

Therefore, restored optical signal can be given as follows:

$$\hat{l}_\lambda(p) = \frac{\left(q_\lambda(p) - \hat{BL}_\lambda(p)\right)}{\hat{t}_\lambda(p)} + \widehat{BL_\lambda}(p). \quad (1.9)$$

Based on the aforementioned theory, Peng and Cosman made a comprehensive summary regarding image information estimation based on DCP method [17], and a restoration method based on image blurriness and light absorption (RBLA) was proposed [15]. Li et al. hierarchically estimated the background light using quad-tree subdivision, and their method of transmission map estimation was characterized by achieving minimum information loss [14]. For a superior color fidelity, Chiang et al. analyzed the wavelength of underwater light, and then compensated it to relieve color distortion [13]. Neural networks have recently been utilized for IFM estimation, e.g., Shin et al. proposed a CNN architecture to estimate the background light and transmission map synchronously [18]. Due to the complex information estimation, the adaptability of IFM-based methods is insufficient.

1.2.3 Visual Restoration Based on Information Fusion

As shown in Figure 1.5, ignoring the IFM, the approach proposed by Ancuti et al. derived weight maps from a degraded image, and the restoration was based on information fusion [19]. In detail, this method first derived two inputs based on the degenerated image, i.e., white balanced version and noise-free version. The former is obtained by illumination estimation while the latter is contrasted. Four weight maps are also derived the degenerated image, including:

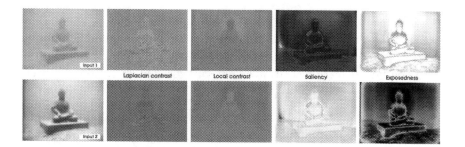

FIGURE 1.5   Fusion-based restoration method [19].

1. Laplacian contrast weight (W_L) constructed with Laplacian filter is used to deal with global contrast, which is sensitive to edges and texture and insensitive to ramps and flat regions.

2. Local contrast weight (W_{LC}) computes the relation among pixel neighborhood. This weight is sensitive to transition regions (e.g., high illumination or shadow), so it is beneficial in describing local contrast.

3. Saliency weight (W_S) pays attention to the loss of object saliency based on saliency detection method. This weight has a drawback that it is sensitive to illumination.

4. Exposedness weight (W_E) is used to moderate the limitation of W_S by estimating over- and under-exposed regions.

After normalization with $\overline{W}^k = W^k \big/ \sum_{k=1}^{K} W^k$, the restored image can be formulated with weighted average as follows:

$$\hat{I}(p) = \sum_{k=1}^{K} \overline{W}^k(p) q^k(p), \tag{1.10}$$

where k indicates input version and $K = 2$.

Compared to IFM-based methods, the method based on information fusion is less affected by the parameters and has higher adaptability. However, it still needs to estimate a large amount of information for each image, and even needs to separately calculate multiple weight maps for multiple derived input images (i.e., white balance version and noise suppression version). Hence, it is difficult for application in real-world scenes for real-time inference.

1.3 REVIEW OF DEEP-LEARNING-BASED OBJECT DETECTION

1.3.1 Two-Stage Detector

The two-stage detection method is also called region-based object detection, and the region can be treated as a candidate proposal. This framework matches the human visual mechanism to a certain extent. First, the entire image is roughly scanned (the generation of candidate frames), and it then focuses on the region of interest for category subdivision and location fine-tuning. Figure 1.6 lists some milestones in the development process of the two-stage detection framework.

1.3.1.1 RCNN

RCNN is the most famous work in the two-stage detection method, proposed by Girshick et al. in 2014 [20]. The subsequent two-stage detection algorithm is improved based on RCNN. As shown in Figure 1.6(a), the training of RCNN contains the following steps:

1. Through region proposal algorithms generate category-independent candidate boxes;

2. Crop all the candidate boxes from the original image and scale them to a uniform size (e.g., 227×227). Use these samples to fine-tune CNN (e.g., AlexNet [5]) to obtain the convolutional features of each candidate box. The CNN needs to be pre-trained on ImageNet [8].

3. Use CNN to extract convolutional features to train a set of class-specific linear support vector machine (SVM) [21] for classification;

4. Use CNN-extracted convolutional features to train category-specific bounding box regression to fine-tune the location of the candidate box.

Despite high detection performance, RCNN has three limitations. First, its offline training process has multiple stages, which are completely individual from each other. Second, its training process needs high computational costs and GPU memory. Third, its inference is not efficient enough. The main reason for the inefficiency is the need to extract features separately for each candidate box, without feature share mechanism.

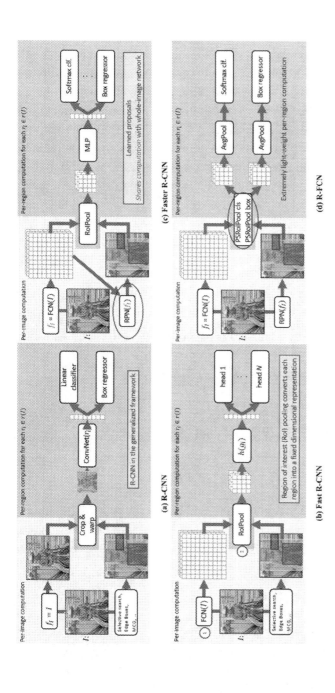

FIGURE 1.6   Two-stage detectors*.

* https://drive.google.com/file/d/1Z1i8mk8Aqdt1_Q8avqtrzLQnLTG7QW6T/view

1.3.1.2 Fast RCNN

Girshick et al. further introduced the multi-task loss of classification and regression and proposed a new CNN detection structure, called Fast RCNN [22]. As shown in Figure 1.6(b), Fast RCNN allows the joint optimization of the network through the multi-task loss, thereby simplifying the training process. Fast RCNN also adopts the mechanism of shared convolutional calculation, and an RoI pooling layer is designed to extract the fixed-length features of each candidate box. It is then input into a series of fully connected layers and split into two branches, i.e., softmax classification and bounding box regression. Compared to RCNN, Fast RCNN greatly improves the efficiency of training and inference; that is, the training speed is 3 times faster and the inference speed is 10 times faster. In short, Fast RCNN can achieve better detection accuracy and learning process.

1.3.1.3 Faster RCNN

Based on Fast RCNN, Ren et al. further proposed Faster RCNN [9]. In the two-stage detection method before Faster RCNN, an extra algorithm is required to predict candidate boxes, which is usually time-consuming. Faster RCNN introduces Region Proposal Network (RPN) to generate candidate boxes. In this manner, RPN and Fast RCNN are unified into a network for joint optimization, and they share backbone features (as shown in Figure 1.6(c)), thus greatly reducing the time cost of candidate box prediction. At the same time, the region proposal generated by RPN will be further regressed and classified for more accurate localization and recall rates. With VGG16 [23] as the backbone, Faster RCNN can achieve a GPU speed of 5 FPS and obtain the best detection performance at that time on PASCAL VOC [6].

1.3.1.4 RFCN

Starting from Fast RCNN, the features of each candidate box are extracted with RoI pooling and then through an RoI-wise subnetwork to further power feature representation. Although the whole process is implemented under the condition of feature share mechanism, the calculation of each candidate box feature in the RoI-wise subnetwork is performed independently. When the subnetwork or the number of candidate box is relatively large, the time-consumption of this part will increase significantly. Therefore, Dai et al. proposed RFCN [24], whose main idea is to minimize the module without sharing calculation. Specifically, RFCN uses convolutional layers to construct the entire network and proposes position-sensitive RoI pooling (PSRoI pooling) instead of using RoI pooling layer. In the last convolutional layer, a series of specific convolutions are introduced to generate position-sensitive feature maps, and then PSRoI pooling is used to extract the features of the candidate boxes. Finally, the weighted voting is performed to obtain the detection confidence and location offset. The structure of RFCN is shown in Figure 1.6(d). Compared to Faster RCNN, RFCN can improve the speed of the detector while achieving comparable detection accuracy.

Based on the two-stage detection method, a series of improvements have also been made. Yang et al. proposed a multi-stage cascading method [25], which cascades a binary Fast R-CNN after region proposal to further filter out some simple backgrounds. The second stage also uses a cascaded Fast R-CNN network for more accurate multi-category prediction by multi-step classification. For dealing with occlusion of objects, Wang et al. proposed an Adversarial Spatial Dropout Network (ASDN) [26], which predicts occlusion mask for each candidate box to simulate the situation of occlusion, and then used the occlusion-aware features for classification and regression. ASDN is analogical to data augmentation at the feature level. For describing the geometric deformation of the object, Dai et al. proposed Deformable Convolutional Networks (DCN) [27]. By learning offsets for each convolution kernel, an irregular filter is constructed to capture the geometrical deformation of the object. In addition, Zhu et al. further proposed DCNv2 [28], which also predicts weights for convolutional kernel to model the importance of different offsets and obtain better ability of deformation. Expanding the two-stage detection method into a multi-task framework, He et al. proposed Mask R-CNN [29], where detection, key points prediction, and instance segmentation can be jointly optimized.

1.3.2 Single-Stage Detector

1.3.2.1 YOLO

Redmon et al. proposed the first single-stage target detection method in the era of deep learning, called YOLO [10]. As shown in Figure 1.7, YOLO divides the image into multiple sub-grids. If the center of the object falls into a grid, the grid is responsible for predicting its category and location. As shown in the bottom of Figure 1.7, the YOLO framework only contains a forward network, producing high computational efficiency. However,

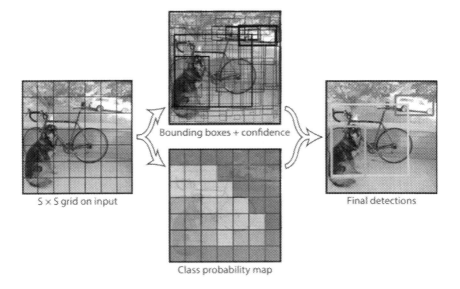

S × S grid on input

Bounding boxes + confidence

Final detections

Class probability map

FIGURE 1.7 YOLO [10].

the framework has two limitations: 1) It cannot flexibly handle different input resolutions owing to the use of fully connected layer and 2) It has poor ability to small object detection. That is, when a grid contains multiple small objects, at most only one can be detected correctly.

Therefore, Redmon et al. proposed YOLOv2 [30], which leveraged anchor mechanism and convolutional prediction layers instead of fully connected prediction. Moreover, anchors are pre-processes using K-means clustering method so that they can highly match objects. In addition, a multi-scale training method is introduced to effectively enhance the robustness for object scale.

On the basis of YOLOv2, Redmon et al. proposed YOLOv3 [31] and introduced a multi-scale prediction mechanism (introduced in Section 2.3.2.2). In addition, logistic regression, instead of traditional softmax, was used for predicting detection confidence. Since softmax can be essentially replaced by multiple independent logistic regression classifiers, where inter-category interference is relatively small, this method is more suitable for multi-label classification.

1.3.2.2 SSD

Liu et al. designed a single-shot multibox detector (SSD) [11] which uses a single-stage network and multi-scale anchors to achieve real-time

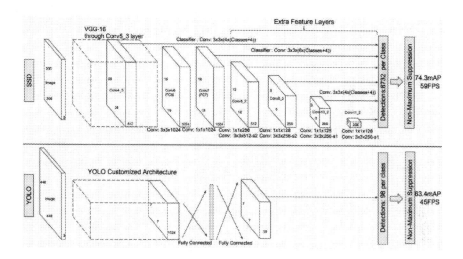

FIGURE 1.8   SSD [11].

high-accuracy object detection. As shown in Figure 1.8, SSD uses a forward backbone network to extract multi-scale features, then predicts the coordinate offset and category confidence of anchors with convolution. NMS is utilized to select the final results among CNN's prediction. In particular, SSD uses multi-scale feature for prediction so that CNN's feature representation can be sufficiently leveraged. In this mechanism, the low-level feature map contains more image details and is helpful for detecting small-scale objects, while the high-level feature map contains more image semantics and is beneficial for detecting large-scale objects.

The key factor to this multi-scale detection lies in the design of the multi-scale anchors, which are designed with multiple scales and aspect ratios. As shown in Figure 1.9, each feature map corresponds to a set of anchors to describe scale-specific objects. For example, the 8×8 convolution feature map with corresponding anchors is fit for detecting "cat", while the 4×4 convolution feature with corresponding anchors are suited to detect "dog". Concretely, CNN's feature maps have different receptive field sizes, so the anchor is designed based on the size of receptive fields. If m feature maps are used for prediction, the scale of the reference frame corresponding to each feature map is:

$$s_k = s_{\min} + \frac{s_{\max} - s_{\min}}{m-1}(k-1), \quad k = 1, 2, \ldots, m, \tag{1.11}$$

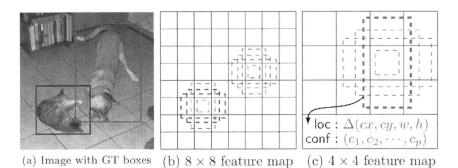

(a) Image with GT boxes   (b) 8×8 feature map   (c) 4×4 feature map

FIGURE 1.9   Multi-scale anchors [11].

where $s_{\min} = 0.2$, $s_{\max} = 0.9$, that is, the anchors corresponding to the bottom feature are 0.2 times the size of input resolution, while the anchors corresponding to the top feature are 0.9 times the size of the input resolution. The aspect ratio of anchors $a_r \in \left\{ 1, 2, 3, \dfrac{1}{2}, \dfrac{1}{3} \right\}$, so the anchor width $w_k^a = s_k \sqrt{a_r}$ and anchor height $h_k^a = s_k / \sqrt{a_r}$. In addition, an anchor with aspect ratio of 1 and size of $\sqrt{s_k s_{k+1}}$ is also designed. Finally, 6 anchors are placed at the feature map cell, whose center point can be expressed as $\left(\dfrac{i_x + 0.5}{|f_k|}, \dfrac{i_y + 0.5}{|f_k|} \right)$, where $|f_k|$ is the size of the kth feature map and $i_x, i_y \in [0, |f_k|)$. At the same time, SSD claims that how to design anchors is a key issue of the detection framework, inspiring subsequent works.

In terms of network training, SSD first classifies anchors into positive and negative samples according to the ground truth, then the training process is based on the multi-task objective function. Let $x_{i,j}^p \in \{1, 0\}$, where 1 means that the i reference frame and the j truth box match on the category p. The multi-task objective function consists of location loss $\mathcal{L}_{loc}$ and classification loss $\mathcal{L}_{conf}$:

$$\mathcal{L}(x, c, l, g) = \frac{1}{N}(\mathcal{L}_{conf}(x, c) + \mathcal{L}_{loc}(x, l, g)), \tag{1.12}$$

where N is the number of positive samples; l is predicted boxes; c denotes category labels; g is ground truth boxes; $\mathcal{L}_{loc}$ is a smooth L1 loss between l and g for cx, cy, w, h:

$$\mathcal{L}_{loc}(x,l,g) = \sum_{i=1}^{N} \sum_{m \in \{cx,cy,w,h\}} x_{i,j}^{k} \text{ smooth L1}(l_i^m - \hat{g}_j^m)$$

$$\hat{g}_j^{cx} = (g_j^{cx} - d_i^{cx})/d_i^w \qquad \hat{g}_j^{cy} = (g_j^{cy} - d_i^{cy})/d_i^h \qquad (1.13)$$

$$\hat{g}_j^w = \log\frac{g_j^w}{d_i^w} \qquad \hat{g}_j^h = \log\frac{g_j^h}{d_i^h}$$

where d_i is the location of anchors; $\mathcal{L}_{conf}$ can be formulated as multi-class cross-entropy:

$$\mathcal{L}_{conf}(x,c) = -\sum_{i=1}^{N} x_{ij}^p \log(c_i^p) - \sum_{i=1}^{\delta N} \log(c_i^0), \qquad (1.14)$$

where $\hat{c}_i^p = \dfrac{\exp(c_i^p)}{\sum_p \exp(c_i^p)}$; 0 denotes background (negative samples); δ is the ratio of positive and negative samples. Because of class imbalance problem, SSD adopts hard negative mining for training, where only high-confidence negative samples are used to optimize and $\delta = 1{:}3$. In addition, SSD designs data augmentation (e.g., random crop and random color distortion) to improve model generalization.

1.3.2.3 RetinaNet

The performance of SSD for detecting small object is not good enough. The reason is that small targets rely on shallow convolution features, and meanwhile the semantic representation of these features is insufficient. Lin et al. leveraged feature pyramid networks (FPN) to the single-stage detection framework, called RetinaNet [32] (as shown in Figure 1.10), which merged high-level features with low-level features to enhance the receptive field of shallow convolutional features. RetinaNet significantly improves the detection performance of single-stage detectors for small targets.

At the same time, Lin et al. believed that the training method based on hard sample mining did not effectively use negative samples for learning, so they designed focal loss to solve the problem of class imbalance. For a sample that matches the category p, the mathematical expression of focal loss is:

$$\text{FL}(\hat{c}^p) = -(1 - \hat{c}^p)^\gamma \log(\hat{c}^p), \qquad (1.15)$$

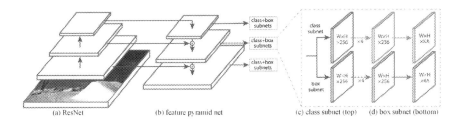

FIGURE 1.10   RetinaNet [32].

where $\hat{c}^p$ is the prediction confidence for the category p. Compared with the traditional cross-entropy loss, focal loss adds a modulation term $(1-\hat{c}^p)^\gamma$ $(\gamma \geq 0)$, which is used to emphasize hard simples. The resulting loss gap is to strengthen the impact of hard samples. In addition, γ can control the intensity of the modulation. Focal loss uses all samples for training and effectively solves the problem of class imbalance. The experimental results show that $\gamma=2$ is the optimal parameter setting.

1.3.2.4 RefineDet

Compared to the two-stage detection method, the single-stage detection method has two limitations: 1) The location accuracy of the single-step regression is relatively low and 2) the category imbalance exists in the single-stage detection training. Zhang et al. proposed RefineDet [33], which introduced a two-step regression into single-stage detection framework and used the negative sample filtering to alleviate the problem of class imbalance. Based on RetinaNet's backbone, RefineDet formulates the down-sampling network as anchor refinement module (ARM), while the feature pyramid network is called object detection module (ODM). As shown in Figure 1.11, ARM features are used to improve anchors and filter simple negative samples. That is, the down-sampled features are sent to a location head and a classification head. The former is used to predict refined anchors for improving regression accuracy, while the latter is used to discard easy negative samples to alleviate class imbalance problems. RefineDet proposes the idea of learning anchors, avoiding the limitations of artificially designed anchors. However, it also causes the problem of anchor–feature mis-alignment between refined anchors and feature sampling location.

1.3.3 Temporal Object Detection

Video information is interrelated across frames; that is, the visual content of adjacent frames is highly related, namely temporal correspondence.

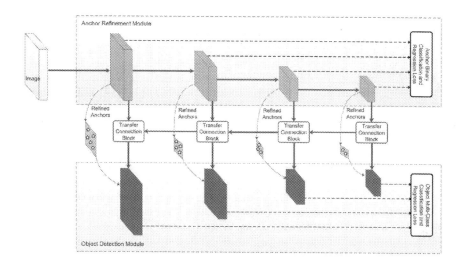

FIGURE 1.11   RefineDet [33].

Existing temporal detection methods use their own approaches to describe temporal correspondence, thereby improving the performance of object detection (i.e., detection accuracy or speed).

1.3.3.1 Post-Processing

Earlier temporal detection methods used temporal correspondence to post-process the results of static detection. For example, inspired by NMS, Han et al. used temporal correspondence to adjust the static detection results of the entire video, thereby proposing the SeqNMS method [34]. SeqNMS has three steps: 1) Sequence selection, i.e., selecting an object with higher confidence in each frame to form an object sequence; 2) Re-scoring, i.e., adjusting the classification confidence of each frame according to the overall confidence of the sequence to enhance temporal correspondence; and 3) Selecting detection results based on intersection over union (IoU). This method makes effective use of temporal information, but its contribution to the detector itself is insufficient (Figure 1.12).

1.3.3.2 Cascade of Detection and Tracking

Single object tracking (SOT) is closely related to object detection, which also needs to locate and classify an object. The difference is that classification in SOT is based on object similarity, while object detection task models semantic information. Therefore, tracking and detection methods can be combined to fuse similarity and semantic information for detection.

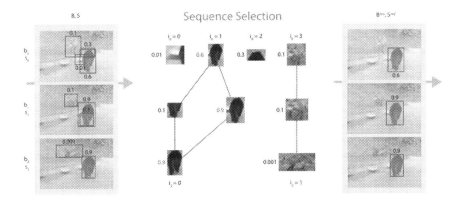

FIGURE 1.12   SeqNMS [34].

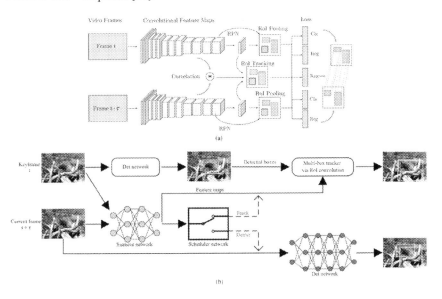

FIGURE 1.13   Cascade of detection and tracking [35, 36].

For example, Feichtenhofer et al. proposed a fused detection and tracking method (D&T) [35] based on a two-stage detector and a correlation filter. As shown in Figure 1.13(a), the similarity measurement from the tracker will be propagated into the detector. This method effectively enhances the classification and location performance of the detector, and can better describe the unclear visual information (e.g., motion blur). As shown in Figure 1.13(b), in order to improve the speed of temporal detection, Luo et al. proposed a parallel detection or tracking framework (DorT) [36].

This method uses a scheduling network to determine whether the current frame should be processed by the detector or the tracker, effectively improving inference speed. It can be seen that cascade of detection and tracking can effectively improve the detection capability, but the model complexity is also relatively high.

1.3.3.3 Feature Fusion Based on Motion Estimation

In the abovementioned method, the tracker actually provides object similarity as the temporal information, but more straightforward temporal information is object motion. Optical flow is a common motion estimation tool, so it is often used in temporal object detection. For example, Zhu et al. claimed that the difference between video and image detection is that low-quality video data may destroy object features. However, this damage is usually short in time, so we can use adjacent features to repair it. The key to aggregate adjacent features is spatial alignment. As shown in Figure 1.14, current frame does not produce effective object description, but the previous and latter frames have good object features, so this method uses optical flow to predict the relative across frames. Thereby, adjacent features are aggregated to the current feature, resulting in a better object description. In addition to optical flow–based motion estimation, Bertasius et al. used deformable convolution to achieve feature alignment and effective feature aggregation [38]. Feature aggregation method based on motion information can construct robust object features across frame, but the time cost is usually high.

1.3.3.4 Feature Propagation Based on RNN

Temporal object detection is a sequence task, which requires algorithms to detect objects in image sequence. Recurrent neural network (RNN) is a typical sequence model, which has been successfully applied to other sequence tasks (e.g., text recognition and speech recognition). Recently, RNN has been utilized in temporal detection tasks. For example, Liu and Zhu inserted long short-term memory (LSTM) modules between visual features and propagated multi-scale features across time [39]. In order to obtain a higher inference speed, a bottleneck LSTM to obtain a better trade-off between accuracy and speed. Compared with feature aggregation based on motion estimation, RNN-based feature propagation has a speed advantage. Moreover, because of the forget gate of LSTM, RNN can select temporal features. However, feature fusion/propagation is

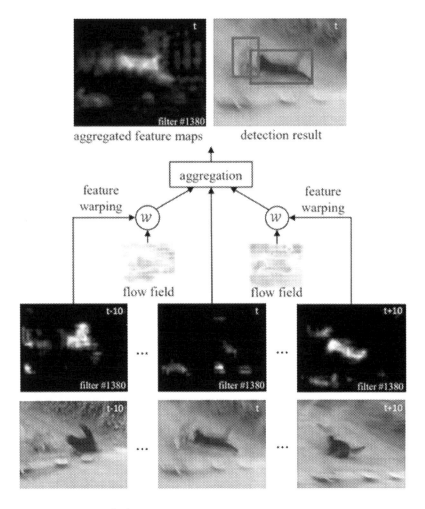

aggregated feature maps   detection result

FIGURE 1.14   FGFA [37].

designed for feature processing, and it does not change the static detection mode.

1.3.3.5 Temporally Sustained Proposal

Based on the temporal correspondence of video, frames can be divided into key frames and non-key frames. Different processing methods for key frames and non-key frames can bring about performance improvements of temporal detection. For example, Chen et al. proposed a detection method based on scale-time lattice (STLattice) to reasonably allocate computing resources [40]. As shown in Figure 1.15, this method includes a feature propagation module and a dynamic key frame selection module, which divide temporal detection into three steps: 1) Static detection is performed on key frames to obtain detection results on sparse key frames; 2) plan a path from the key frames to non-key frames; and 3) based on this path, propagate and refine the detection results at each time step. In this manner, time efficiency can be improved since the number of key frames is much less than the non-key frames. Moreover, different accuracy and speed trade-offs can be obtained by adjusting the sparseness of key frames.

1.3.3.6 Batch-Processing

Temporal correspondence of video data allows the detector to analyze multi-frame features at the same time, and then conduct object classification and regression for multiple frames. For example, inspired by the region proposal network (RPN), Kang et al. proposed a tubelet proposal network (TPN), where temporally propagated detection boxes form object tubelets. As shown in Figure 1.16, TPN first set up anchor tubule for multiple frames, then used LSTM to classify and localize them. This method can produce stable temporal detection results, but this offline processing method is difficult to apply to real-world tasks.

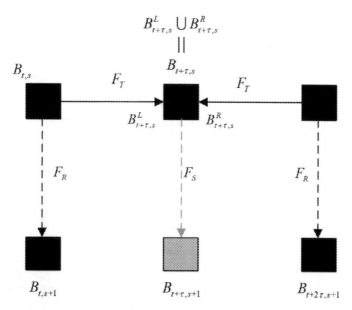

FIGURE 1.15   STLattice [40].

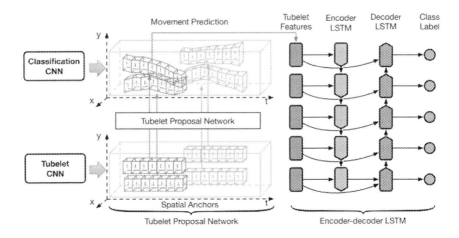

FIGURE 1.16   TPN [41].

1.3.4 Benchmarks of Object Detection

1.3.4.1 PASCAL VOC

PASCAL VOC is one of the representative datasets in the field of object detection [6]. This dataset is divided into two sub-datasets, i.e., VOC2007 and VOC2012, which require the algorithm to predict objects among 20 categories (i.e., vehicles, airplanes, trains, cats, dogs, and birds). VOC2007 contains 9,963 image samples, which are divided into training set, validation set, and test set, which contain 2,501, 2,510, and 4,952 image samples, respectively. VOC2012 contains 22,531 image samples, which are also divided into the training set, the validation set and the test set, which contain 5,717, 5,823, and 1,091 samples, respectively.

In general, as for VOC2007 setting, the training set and the validation set of VOC2007 and VOC2012 (called *07+12*, a total of 16,551 image samples) are combined for training, and the VOC2007 test set is used for evaluation. In terms of VOC2012, the training set and the validation set of VOC2007 and VOC2012 as well as the VOC2007 test set (called *07++12*, a total of 21,503 image samples) are combined for training, and then the VOC2012 test set is leveraged for evaluation.

1.3.4.2 MS COCO

MS COCO is a dataset for object detection and segmentation tasks in recent years [7]. It is also one of the most popular data sets in the scope of object detection. The dataset requires the algorithm to predict objects among 80 categories (e.g., people, bottles, and chairs.). In 2014, the dataset disclosed 82,783 training samples, 40,504 validation samples, and 40,775 test samples. Although there were adjustments later, the training set and the validation set were almost unchanged. The annotation of the training set and the validation set are available, and the annotations of the test set are unpublished. One need to submit results online for evaluation. In addition, MS COCO divides objects into three types according to object size. That is, small, middle, and large objects should have object sizes of $<32^2$, $<96^2$, and $>96^2$, respectively.

Usually only 5,000 samples in the validation set are used for validation (called *minival2014*), while the training set and the remaining validation set (called *trainval35k*, a total of 118287 image samples) are combined for training. Evaluation process is based on the test set (called *test-dev*, a total of 20288 image samples).

1.3.4.3 ImageNet VID

Unlike PASCAL VOC and MS COCO, samples of the ImageNet VID dataset are videos. This dataset requires the algorithm to predict objects among 30 categories (e.g., sheep, dog, airplane, ship, etc.) in videos. The training set of ImageNet VID contains 4,000 video samples, and the validation set contains 1,314 video samples.

Although there are abundant image data, the disadvantage of video data is the lack of diversity, which damages model generalization. Therefore, the ImageNet DET dataset is usually used to train static detectors. DET contains 200 object categories and the categories of VID is a subset of that DET, so the combined VID and DET (only VID-consistent categories) are used for training. Specifically, at most 2,000 samples are selected for each category from DET and 10 samples are selected for each category from VID to form the training set. Model evaluation is conducted based on the validation set of VID.

1.3.4.4 Evaluation Metrics

Object detection is evaluated with an average precision (AP). First, the IoU between prediction and ground truth is calculated, and thus prediction is divided into 3 categories.

1) True positive (TP): Prediction is an object, and it matches a ground truth.

2) False positive (FP): Prediction is an object, but it does not match a ground truth

3) False negative (FN): Under the condition of specific IoU, there is no prediction that can match the ground truth.

Therefore, recall rate can be formulated as:

$$R = \frac{TP}{TP + FN}.$$ (1.16)

Precision can be formulated as:

$$P = \frac{TP}{TP + FP}.$$ (1.17)

According to detection confidence, detection results can be sorted. Thereby, different recall rates can be obtained based on confidence threshold, and P can be described as a function of R so that the P-R curve can be obtained. As a result, AP is the area under P-R curve as follows:

$$AP = \int_0^1 P(R)\,dR.$$ (1.18)

AP over multiple classes, i.e., mean average precision (mAP) can be formulated as:

$$mAP = \frac{1}{C}\sum_{i=1}^{C} AP_i.$$ (1.19)

In most benchmarks, the IoU is fixed, e.g., 0.5 in PASCAL VOC. In contrast, evaluation of MS COCO is based on uniformly varying IoU, i.e., IoU $\in$ [0.5:0.05:0.95]. In this manner, 10 mAP are obtained, and their average is used to describe detection performance. In addition, MS COCO evaluates size-related AP, producing AP_S, AP_M, and AP_L for small, middle, and large objects, respectively.

1.4 REVIEW OF UNDERWATER STEREO MEASUREMENT

In a limited range of circumstances, calibration might not be necessary. If a high level of accuracy is not required, and the object to be measured approximates a two-dimensional planar surface, a very straightforward solution is possible. Correction lenses or dome ports [42, 43] are used to provide near-perfect central projection underwater by eliminating the refraction effects. The remaining small errors can be either corrected using a grid or graticule placed in the field of view or accepted as a small deterioration in accuracy. The primary advantage of the correction lens or dome port is that there is little degradation of image quality near the edges of the port.

However, there are numerous application fields that need a high level of accuracy. Therefore, if accuracy is a priority, and especially if the object to be measured is a three-dimensional surface, then a comprehensive calibration is essential. The alternative approach of a comprehensive calibration translates a reliable technique from air into the underwater environment. Close-range calibration of cameras is a well-established technique that was proposed by [44], and it was extended to self-calibration [45] and adapted to the underwater environment [46, 47]. The mathematical basis of the technique is detailed in [48].

There are two primary reasons for the necessity of calibration for a camera system.

First, the internal geometric characteristics of the cameras must be determined [44]. In photogrammetric practice, camera calibration is most often defined by a physical parameter set comprising principal distance, principal point location, radial [49] and decentering [50] lens distortions, plus affinity and orthogonality terms to compensate for minor optical effects [51, 52]. The principal distance is formally defined as the separation, along the camera optical axis, between the lens-perspective center and the image plane. The principal point is the intersection of the camera optical axis with the image plane.

Second, the relative orientation of the cameras with respect to one another, or the exterior orientation with respect to an external reference, must be determined. Also known as pose estimation, both the location and orientation of the camera(s) must be determined. For the commonly used approach of stereo cameras, the relative orientation effectively defines the separation of the perspective centers of the two lenses, the pointing angles (omega and phi rotations) of the two optical axes of the cameras and the roll angles (kappa rotations) of the two focal plane sensors.

In the underwater environment the effects of refraction must be corrected or modeled to obtain an accurate calibration. The entire light path, including the camera lens, housing port, and water medium, must be considered. By far the most common approach is to correct the refraction effects using absorption by the physical camera calibration parameters. Assuming that the camera optical axis is approximately perpendicular to a plane or dome camera port, the primary effect of refraction through the air-port and port-water interfaces will be radially symmetric around the principal point [53]. This primary effect can be absorbed by the radial lens distortion component of the calibration parameters. There will also be some small, asymmetric effects caused by, for example, alignment errors between the optical axis and the housing port, and perhaps non-uniformities in the thickness or material of the housing. These secondary effects can be absorbed by calibration parameters such as the decentering lens distortion and the affinity term.

The disadvantage of the absorption approach for the refractive effects is that there will always be some systematic errors which are not incorporated into the model. The effect of refraction invalidates the assumption of a single projection center for the camera [54], which is the basis for the physical parameter model. The errors are most often manifested as scale changes when measurements are taken outside of the range used for the calibration process. Experience over many years of operation demonstrates that if the ranges for the calibration and the measurements are commensurate, then the level of systematic error is generally less than the precision with which measurements can be extracted. This masking effect is partly due to the elevated level of noise in the measurements, caused by the attenuation and loss of contrast in the water medium.

The alternative to the simple approach of absorption is the more complex process of geometric correction, effectively an application of ray tracing of the light paths through the refractive interfaces. A two-phase approach is developed in [55] for a stereo camera housing with concave lens covers. An in-air calibration is carried out first, followed by an in-water calibration that introduces 11 lens cover parameters such as the center of curvature of the concave lens and, if not known from external measurements, refractive indices for the lens covers and water. A more general geometric correction solution is developed for plane port housings in [56]. Additional unknowns in the solution are the distance between the camera perspective center and the housing, and the normal of the plane housing port, whilst the port thickness and refractive indices must be known. Using ray tracing, [57] developed a general solution to refractive surfaces that, in theory, can accommodate any shape of camera housing port. The shape of the refractive surface and the refractive indices must be known.

A variation on the geometric correction is the perspective center shift or virtual projection center approach. A specific solution for a planar housing port is developed in [58]. The parameters include the standard physical parameters, the refractive indices of glass and water, the distance between the perspective center and the port, the tilt and direction of the optical axis with respect to the normal to the port, and the housing interface thickness. A modified approach neglects the direction of the optical axis and the thickness of thin ports, as these factors can be readily absorbed by the standard physical parameters. Again, a two-phase process is required, first a "dry" calibration in air and then a "wet" calibration in water [58]. A similar principle is used in [59], also with a two-phase calibration approach.

The advantage of these techniques is that, without the approximations in the models, the correction of the refractive effects is exact. The disadvantages are the requirements for two-phase calibrations and known data such as refractive indices. Further, in some cases, the theoretical solution is specific to a housing type, whereas the absorption approach has the distinct advantage that it can be used with any type of underwater housing.

A review of refraction correction methods for underwater imaging is given in [54]. The perspective camera model, ray-based models and physical models are analyzed, including an error analysis based on synthetic data. The analysis demonstrates that perspective camera models incur increasing errors with increasing distance and tilt of the refractive surfaces, and only the physical model of refraction correction permits a complete theoretical compensation.

Once the camera calibration is established, single camera systems can be used to acquire measurements when used in conjunction with reference frames [60] or sea floor reference marks [61]. For multi-camera systems the relative orientation is required as well as the camera calibration. The relative orientation can be included in the self-calibration solution as a constraint [62] or can be computed as a post-process based on the camera positions and orientations for each set of synchronized exposures [47]. In either case, it is important to detect and eliminate outliers, usually caused by lack of synchronization, that would otherwise unduly influence the calibration solution or the relative orientation computation. Outliers caused by synchronization effects are more common for systems based on camcorders or video cameras in separate housings, which typically use an external device such as a flashing LED light to synchronize the images to within one frame [47].

With an efficient calibration algorithm for the underwater stereo measurement, the accuracy of measurement will be improved significantly. In this context, the applications of underwater robots will benefit from the reliable measurement technologies.

1.5 OVERVIEW OF THE SUBSEQUENCE CHAPTERS

In Chapter 2, a novel deep-learning-based underwater image restoration method and a filter-based underwater image restoration method are introduced. Chapter 3 focuses on an innovative calibration algorithm underwater stereo measurement. The object detection strategies with various frameworks of neural network are described in detail in Chapters 4–6. Two projects of vision-based underwater robotic application, including an intelligent surface cleaning robot and a bio-inspired fish for vision-based target tracking, are exhibited in Chapters 7 and 8, which will help readers comprehend the underwater robotic vision and motion control much more further.

REFERENCES

1. J. Yuan, Z. Wu, J. Yu, C. Zhou, and M. Tan, "Design and 3D motion modeling of a 300-m gliding robotic dolphin," in *Proceedings of the World Congress of the International Federation of Automatic Control*, Toulouse, France, Jul. 2017, pp. 12685–12690.
2. M. Cai, Y. Wang, S. Wang, R. Wang, Y. Ren, and M. Tan, "Grasping marine products with hybrid-driven underwater vehicle-manipulator system," *IEEE Trans. Autom. Sci. Eng.*, vol. 17, no. 3, pp. 1443–1454, 2020, doi: 10.1109/TASE.2019.2957782.
3. Z. Gong, J. Cheng, X. Chen, W. Sun, X. Fang, K. Hu, Z. Xie, T. Wang, and L. Wen, "A bioinspired soft robotic arm: Kinematic modeling and hydrodynamic experiments," *J. Bionic Eng.*, vol. 15, no. 2, pp. 204–219, 2018.
4. Z. Zhao, P. Zheng, S. Xu, and X. Wu, "Object detection with deep learning: A review," *IEEE Trans. Neural Networks Learn. Syst.*, vol. 30, no. 11, pp. 3212–3232, 2019.
5. Krizhevsky, I. Sutskever, and G. E. Hinton, "ImageNet classification with deep convolutional neural networks," in *Proceedings of the Advances in Neural Information Processing Systems*, Lake Tahoe, USA, Jun. 2012, pp. 1097–1105.

6. M. Everingham, L. Van Gool, C. K. Williams, J. Winn, and A. Zisserman, "The pascal visual object classes (VOC) challenge," *Int. J. Comput. Vis.*, vol. 88, no. 2, pp. 303–338, 2010.

7. T. Y. Lin, M. Maire, S. Belongie, J. Hays, P. Perona, D. Ramanan, P. Dollár, and C. L. Zitnick, "Microsoft COCO: Common objects in context," in *Proceedings of the European Conference on Computer Vision*, Zurich, Switzerland, Sep. 2014, pp. 740–755.

8. O. Russakovsky, J. Deng, H. Su, J. Krause, S. Satheesh, S. Ma, Z. Huang, A. Karpathy, A. Khosla, M. Bernstein, A. C. Berg, and F. Li, "ImageNet large scale visual recognition challenge," *Int. J. Comput. Vis.*, vol. 115, no. 3, pp. 211–252, 2015.

9. S. Ren, K. He, R. Girshick, and J. Sun, "Faster R-CNN: Towards real-time object detection with region proposal networks," in *Proceedings of the Advances in Neural Information Processing Systems*, Montreal, Canada, Dec. 2015, pp. 91–99.

10. J. Redmon, S. S. Divvala, R. Girshick, and A. Farhadi, "You only look once: Unified, realtime object detection," in *Proceedings of the IEEE Conference on Computer Vision and Pattern Recognition*, Las Vegas, USA, Jun. 2016, pp. 779–788.

11. W. Liu, D. Anguelov, D. Erhan, C. Szegedy, S. Reed, C. Y. Fu, and A. C. Berg, "SSD: Single shot multibox detector," in *Proceedings of the European Conference on Computer Vision*, Amsterdam, Netherlands, Oct. 2016, pp. 21–37.

12. Y. Y. Schechner and N. Karpel, "Clear underwater vision," in *Proceedings of the IEEE Conference on Computer Vision and Pattern Recognition*, Washington, USA, Jun. 2004, pp. I-536–I-543.

13. J.-Y. Chiang and Y. Chen, "Underwater image enhancement by wavelength compensation and dehazing," *IEEE Trans. Image Process.*, vol. 21, no. 4, pp. 1756–1769, 2012.

14. C. Li, J. Guo, R. Cong, Y. Pang, and B. Wang, "Underwater image enhancement by dehazing with minimum information loss and histogram distribution prior," *IEEE Trans. Image Process.*, vol. 25, no. 12, pp. 5664–5677, 2016.

15. Y.-T. Peng and P. C. Cosman, "Underwater image restoration based on image blurriness and light absorption," *IEEE Trans. Image Process.*, vol. 26, no. 4, pp. 1579–1594, 2017.

16. S. Emberton, L. Chittka, and A. Cavallaro, "Hierarchical rank-based veiling light estimation for underwater dehazing," in *Proceedings of the British Machine Vision Conference*, Swansea, UK, Sep. 2015, pp. 125.1–125.12.

17. K. He, J. Sun, and X. Tang, "Single image haze removal using dark channel prior," *IEEE Trans. Pattern Anal. Mach. Intell.*, vol. 33, no. 12, pp. 2341–2353, 2011.

18. Y.-S. Shin, Y. Cho, G. Pandey, and A. Kim, "Estimation of ambient light and transmission map with common convolutional architecture," in *Proceedings of the IEEE/MTS OCEANS*, Monterey, USA, Sep. 2016, pp. 1–7.

19. C. Ancuti, C. O. Ancuti, T. Haber, and P. Bekaert, "Enhancing underwater images and videos by fusion," in *Proceedings of the IEEE Conference on Computer Vision and Pattern Recognition*, Providence, Rhode Island, Jun. 2012, pp. 81–88.

20. R. Girshick, J. Donahue, T. Darrell, and J. Malik, "Rich feature hierarchies for accurate object detection and semantic segmentation," in *Proceedings of the IEEE Conference on Computer Vision and Pattern Recognition*, Columbus, USA, Jun. 2014, pp. 580–587.

21. M. A. Hearst, S. T. Dumais, E. Osuna, J. Platt, and B. Scholkopf, "Support vector machines," *IEEE Intell. Syst. Applications*, vol. 13, no. 4, pp. 18–28, 1998.

22. R. Girshick. "Fast R-CNN," in *Proceedings of the IEEE International Conference on Computer Vision*, Santiago, Chile, Dec. 2015, pp. 1440–1448.

23. K. Simonyan and A. Zisserman, "Very deep convolutional networks for large-scale image recognition," https://arxiv.org/abs/1409.1556, 2014.

24. J. Dai, Y. Li, K. He, and J. Sun, "R-FCN: Object detection via region-based fully convolutional networks," in *Proceedings of the Advances in Neural Information Processing Systems*, Barcelona, Spain, Dec. 2016, pp. 379–387.

25. B, Yang, J. Yan, Z. Lei, and S. Z. Li, "Craft objects from images," in *Proceedings of the IEEE Conference on Computer Vision and Pattern Recognition*, Las Vegas, USA, Jun. 2016, pp. 6043–6051.

26. X. Wang, A. Shrivastava, and A. Gupta, "A-Fast-RCNN: Hard positive generation via adversary for object detection," in *Proceedings of the IEEE Conference on Computer Vision and Pattern Recognition*, Hawaii, USA, Jun. 2017, pp. 2606–2615.

27. J. Dai, H. Qi, Y. Xiong, Y. Li, G. Zhang, H. Hu, and Y. Wei, "Deformable convolutional networks," in *Proceedings of the IEEE International Conference on Computer Vision*, Venice, Italy, Oct. 2017, pp. 764–773.

28. X. Zhu, H. Hu, S. Lin, J. Dai, "Deformable ConvNets v2: More deformable, better results," in *Proc. Conference on Computer Vision and Pattern Recognition (CVPR)*, California, USA, 2019, pp. 9308–9316. *arXiv: 1811.11168*, 2018.

29. K. He, G. Gkioxari, P. Dollár, and R. Girshick, "Mask R-CNN," in *Proceedings of the IEEE International Conference on Computer Vision*, Venice, Italy, Oct. 2017, pp. 2961–2969.

30. J. Redmon and A. Farhadi, "YOLO9000: Better, faster, stronger," in *Proceedings of the IEEE Conference on Computer Vision and Pattern Recognition*, Honolulu, USA, Jun. 2017, pp. 6517–6525.

31. J. Redmon and A. Farhadi, "YOLOv3: An incremental improvement," https://arxiv.org/abs/1804.02767, 2018.

32. T. Y. Lin, P. Goyal, R. Girshick, K. He, and P. Dollár, "Focal loss for dense object detection," in *Proceedings of the IEEE International Conference on Computer Vision*, Venice, Italy, Oct. 2017, pp. 2980–2988.

33. S. Zhang, L. Wen, X. Bian, Z. Lei, and S. Z. Li, "Single-shot refinement neural network for object detection," in *Proceedings of the IEEE Conference on*

Computer Vision and Pattern Recognition, Salt Lake City, USA, Jun. 2018, pp. 4203–4212.

34. W. Han, P. Khorrami, T. L. Paine, P. Ramachandran, M. Babaeizadeh, H. Shi, J. Li, S. Yan, and T. S. Huang, "Seq-NMS for video object detection," https://arxiv.org/abs/1602.08465, 2016.

35. C. Feichtenhofer, A. Pinz, and A. Zisserman, "Detect to track and track to detect," in *Proceedings of the IEEE International Conference on Computer Vision*, Venice, Italy, Oct. 2017, pp. 3038–3046.

36. H. Luo, W. Xie, X. Wang, and W. Zeng, "Detect or track: Towards cost-effective video object detection/tracking," in *Proceedings of the AAAI Conference on Artificial Intelligence*, Honolulu, USA, Jan. 2019, pp. 8803–8810.

37. X. Zhu, Y. Wang, J. Dai, L. Yuan, and Y. Wei, "Flow-guided feature aggregation for video object detection," in *Proceedings of the IEEE International Conference on Computer Vision*, Venice, Italy, Oct. 2017, pp. 408–417.

38. L. Bertasius, G Torresani and J. Shi, "Object detection in video with spatiotemporal sampling networks," in *Proceedings of the IEEE European Conference on Computer Vision*, Munich, Germany, Sep. 2018, pp. 342–357.

39. M. Liu and M. Zhu, "Mobile video object detection with temporally-aware feature maps," in *Proceedings of the IEEE Conference on Computer Vision and Pattern Recognition*, Salt Lake City, USA, Jun. 2018, pp. 5686–5695.

40. K. Chen, J. Wang, S. Yang, X. Zhang, Y. Xiong, C. C. Loy, and D. Lin, "Optimizing video object detection via a scale-time lattice," in *Proceedings of the IEEE Conference on Computer Vision and Pattern Recognition*, Salt Lake City, USA, Jun. 2018, pp. 7814–7823.

41. K. Kang, H. Li, T. Xiao, W. Ouyang, J. Yan, X. Liu, and X. Wang, "Object detection in videos with tubelet proposal networks," in *Proceedings of the IEEE Conference on Computer Vision and Pattern Recognition*, Honolulu, USA, Jun. 2017, pp. 727–735.

42. E. J. Moore, "Underwater photogrammetry," *Photogramm. Rec.*, vol. 5, pp. 748–763, 1976.

43. Ivanoff and P. Cherney, "Correcting lenses for underwater use," *J. Soc. Motion Pict. Telev. Eng.*, vol. 69, pp. 264–266, 1960.

44. D. C. Brown, "Close range camera calibration," *Photogramm. Eng.*, vol. 37 pp. 855–866, 1971.

45. J. F. Kenefick, M. S. Gyer, and B. F. Harp, "Analytical self calibration," *Photogramm. Eng. Remote Sens.*, vol. 38, pp. 1117–1126, 1972.

46. J. G. Fryer and C. S. Fraser, "On the calibration of underwater cameras," *Photogramm. Rec.*, vol. 12, pp. 73–85, 1986.

47. E. S. Harvey and M. R. Shortis, "A system for stereo-video measurement of sub-tidal organisms," *Mar. Technol. Soc. J.*, vol. 29, pp. 10–22, 1996.

48. S. I. Granshaw, "Bundle adjustment methods in engineering photogrammetry," *Photogramm. Rec.*, vol. 10, pp. 181–207, 1980.

49. M. R. Shortis and J. W. Seager, "A practical target recognition system for close range photogrammetry," *Photogramm. Rec.*, vol. 29, pp. 337–355, 2014.

50. H. Ziemann and S. F. El-Hakim, "On the definition of lens distortion reference data with odd-powered polynomials," *Can. Surv.*, vol. 37, pp. 135–143, 1983.
51. D. C. Brown, "Decentring distortion of lenses," *Photogramm. Eng.*, vol. 22, pp. 444–462, 1966.
52. C. S. Fraser, M. R Shortis, and G Ganci, "Multi-sensor system self-calibration," *Proc. SPIE*, vol. 2598 pp. 2–18, 1995.
53. M. R. Shortis, "Multi-lens, multi-camera calibration of Sony Alpha NEX 5 digital cameras," In *Proceedings of the CD-ROM, GSR_2 Geospatial Science Research Symposium*, Melbourne, Australia, Dec. 2012.
54. Sedlazeck and R. Koch, "Perspective and non-perspective camera models in underwater imaging-Overview and error analysis," In *Outdoor and Large-Scale Real-World Scene Analysis*, F. Dellaert, J.-M. Frahm, M. Pollefeys, L. Leal-Taixé, B. Rosenhahn, Eds. Berlin/Heidelberg, Germany: Springer, 2012, pp. 212–242.
55. R. Li, H. Li, W. Zou, R. G. Smith, and T. A. Curran, "Quantitative photogrammetric analysis of digital underwater video imagery," *IEEE J. Ocean. Eng.*, vol. 22, pp. 364–375, 1997.
56. Jordt-Sedlazeck and R. Koch, "Refractive calibration of underwater cameras," In *Computer Vision-CCV 2012, 12th European Conference on Computer Vision*, Firesnze, Italy. Berlin/Heidelberg, Germany: Springer, 2012; pp. 846–859.
57. R. Kotowski, "Phototriangulation in multi-media photogrammetry," *Int. Arch. Photogramm. Remote Sens.*, vol. 27, pp. 324–334, 1988.
58. G. Telem and S. Filin, "Photogrammetric modeling of underwater environments," *ISPRS J. Photogramm. Remote Sens.*, vol. 65, pp. 433–444, 2010.
59. C. Bräuer-Burchardt, P. Kühmstedt, and G. Notni, "Combination of air- and water-calibration for a fringe projection based underwater 3D-scanner," in *Computer Analysis of Images and Patterns*, G. Azzopardi, N. Petkov, Eds. Cham, Switzerland: Springer, 2015, pp. 49–60.
60. G. F. Bass and D. M. Rosencrantz, "The ASHREAH—A pioneer in search of the past," In *Submersibles and Their Use in Oceanography and Ocean Engineering*, R.A. Geyer, Ed. Amsterdam, The Netherlands: Elsevier, 1977, pp. 335–350.
61. J. Green, S. Matthews, and T. Turanli, "Underwater archaeological surveying using Photomodeler, VirtualMapper: Different applications for different problems," *Int. J. Naut. Archaeol.*, vol. 31, pp. 283–292, 2002.
62. B. R. King, "Bundle adjustment of constrained stereo pairs-Mathematical models," *Geomat. Res. Australas.*, vol. 63, pp. 67–92, 1995.

CHAPTER 2

Adaptive Real-Time Underwater Visual Restoration with Adversarial Critical Learning

2.1 INTRODUCTION

With the rapid development of computer vision and convolutional networks (CNN), a multitude of underwater vision tasks have emerged. For example, overcoming the problem with low-contrast visualization, Chuang et al. tracked live fish with a segmentation algorithm [1]. Additionally, Chen et al. proposed an identity-aware detection method based on single-shot detector (SSD) for underwater object grasping [2, 3]. However, the underwater vision is severely degraded [4], and thus, it is imperative to elevate visual quality for aquatic robots. To that end, some studies on underwater image enhancement have been conducted [5–10]. Nevertheless, the visual degeneration is multifarious (see Figure 2.1), and most of the existing literature has difficulty when dealing with a variety of types of underwater environments using constant parameter settings [7]. Moreover, the problem with low time efficiency is rarely tackled, which is

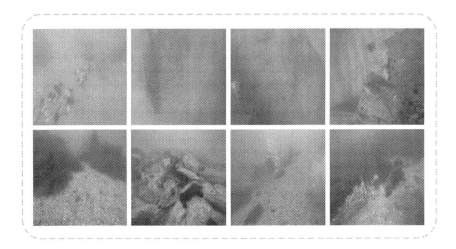

FIGURE 2.1 Various undersea images. Most existing algorithms restore them with complex information estimation and high time costs, whereas we treat this task as a computationally efficient image-to-image translation.

pivotal for robots' autonomous operations. Thus, it is essential to develop a real-time and adaptive method for underwater visual restoration.

Recently, Generative Adversarial Networks (GAN) [11] have been successfully employed in image-to-image translation tasks, e.g., style transfers and super-resolution [12]. It is clear that image restoration can be treated as an image-to-image translation, so we are certain that GAN is able to restore the underwater scenes if trained with paired data (i.e., original underwater images and corresponding in-air versions). Furthermore, a well-trained GAN-based method can adaptively work for various underwater scenarios. When it comes to underwater training data, although paired images are hard to be obtained, synthetic in-air data based on a traditional method can provide unambiguous visual content for training. However, the characteristics of synthetic samples and real in-air data are still distinct to some extent, so synthetic images cannot be employed as the ground truth. Otherwise, GAN's results can perform similarly but no better than the synthetic data. That is, underwater noise that incurs color distortion, contrast decrease, and haziness still needs to be further removed. Thereby, a new framework is required for further enhancement.

In this chapter, to adaptively restore underwater visual quality in real time, we propose an adversarial critical learning (ACL) and a GAN-based restoration scheme (GAN-RS). The tasks of our generator are twofold: (1) Preserving image content and (2) removing underwater noise. To these ends, we build a multibranch discriminator including an adversarial branch and a critic branch. Making image content unambiguous, [13] provides supervision to train the adversarial branch in a supervised manner for the first task. Additionally, a novel dark channel prior (DCP) loss is developed for the same purpose. The DCP loss promotes the pixel-level similarity, whereas the adversarial training is responsible for high-level analogy in terms of features. For the second task, we investigate a creative evaluation criterion for underwater properties, namely, underwater index. Subsequently, the corresponding underwater index loss is designed to train the critic branch, and the combination of the loss functions obeys a multistage loss strategy. Extensive comparison experiments verify the restoration quality, time efficiency, and adaptability of the proposed algorithm. The contributions made in this chapter are summarized as follows:

- We propose a GAN-RS to elevate underwater visual quality. After training, the GAN-RS no longer needs any prior knowledge that makes it work.

- The GAN-RS is learned based on our proposed ACL. In ACL, multibranch discriminator is developed, where an adversarial branch is leveraged for preserving image content while a critic branch is explicitly designed for removing underwater noise. A DCP loss, an underwater index, and a multistage loss strategy are investigated to assure effective training.

- The GAN-RS reaches over 100 frames per second (FPS) and achieves a superior restoration performance.

The rest of the chapter is organized as follows. Section 2.2 provides a detailed review of underwater visual restoration and image-to-image translation. Section 2.3 presents the proposed ACL learning method and GAN-RS restoration framework. Section 2.4 shows the experimental results followed by a discussion of the pros and cons of our proposed method. Section 2.5 concludes the chapter with an outline of future work.

2.2 REVIEW OF VISUAL RESTORATION AND IMAGE-TO-IMAGE TRANSLATION

2.2.1 Traditional Underwater Image Restoration Methods

Most existing methods for restoring underwater images are based on an image formation model (IFM) [5–8], where the background light and transmission map should be estimated in advance. IFM can be formulated as

$$q_\lambda(p) = l_\lambda(p)t_\lambda(p) + BL_\lambda(1 - t_\lambda(p)), \quad \lambda \in \{R, G, B\}, \tag{2.1}$$

where p is pixel position and $\lambda \in \{R, G, B\}$ denotes wavelength; q_λ is degenerated optical signal; l_λ is original optical signal; BL_λ denotes background light (BL); t_λ is transmission map (TM), denoting the ratio of optical decay; t_λ is usually formulated as an exponential decay, i.e., $t_\lambda = e^{-\eta_\lambda} di$, where η is attenuation coefficient; di is object-to-camera distance. Based on the aforementioned theory, Peng and Cosman made a comprehensive summary regarding image information estimation based on DCP method [14], and a restoration method based on image blurriness and light absorption (RBLA) was proposed [7]. Li et al. hierarchically estimated the background light using quad-tree subdivision, and their method of transmission map estimation was characterized by achieving minimum information loss [6]. For a superior color fidelity, Chiang et al. analyzed the wavelength of underwater light and then compensated it to relieve color distortion [5]. Neural networks have recently been utilized for IFM estimation, e.g., Shin et al. proposed a CNN architecture to estimate the background light and transmission map synchronously [15].

On the other hand, ignoring the IFM, the approach proposed by Ancuti et al. derived weight maps from a degraded image, and the restoration was based on information fusion [9]. In detail, this method firstly derives two inputs based on the degenerated image, i.e., white-balanced version and noise-free version. The former is obtained by illumination estimation while the latter is contrasted. Four weight maps are also derived from the degenerated image, including Laplacian contrast weight (W_L), local contrast weight (W_{LC}), saliency weight (W_S), and exposedness weight (W_E). Finally, the original signal can be restored with input versions and weights.

The information estimation in both IFM-based and fusion-based restoration can potentially be a waste of time, so we directly treat the restoration as an image-to-image translation task. The quality of underwater vision can be enhanced by a single-short network in the GAN-RS.

2.2.2 Image-to-Image Translation

With the development of deep learning, particularly GAN [11], approaches to image-to-image translation have been rapidly developed in recent years for *Labels to Street scene, Aerial photo to Map, Day to Night, Edges to Photo*, and so on. If there are paired data, GAN can be trained in a supervised way [12, 16–18]. Zhu et al. used GAN to learn the manifold of natural images, whose generator presented the scenes or objects from the profiles [16]. Combining an adversarial loss with the mean squared error, Ledig et al. constructed a perceptual loss to guide the generator more effectively [17]. Meanwhile, residual blocks were employed to design a generative network. Isola et al. proposed a general framework for supervised image-to-image translation problems based on conditional GAN (cGAN) [19], namely pix2pix, and built a fully convolutional discriminator to concern image patches [18].

In most cases, paired images are hard to obtain, so several unsupervised methods have been developed [20–22]. Dong et al. designed an unsupervised framework with three stages, i.e., learning the shared features, learning the image encoder, and translation [20]. Extending from the pix2pix, Zhu et al. proposed a general unsupervised framework, namely, CycleGAN, whose main idea was the minimization of reconstruction error between two sets of training data [21]. Liu et al. proposed unsupervised image-to-image translation networks based on shared-latent space assumption, where images could be recovered from latent codes [22].

To preserve the image content, supervised methods are more suitable for underwater restoration, and we proposed ACL for learning. That is, despite the paired training data, the target image serves as the supervision of adversarial branch rather than the final ground truth. Moreover, critical learning simultaneously works for guiding image generation.

2.3 GAN-BASED RESTORATION WITH ADVERSARIAL CRITICAL LEARNING

In this section, a filtering-based restoration scheme (FRS) is briefed at the beginning, and the FRS generates real adversarial samples in the GAN training process. Then, the architecture of the proposed GAN-RS for underwater image restoration is detailed, followed by the loss function for training.

2.3.1 Filtering-Based Restoration Scheme

In general, the training process of GAN requires real and fake samples, so we introduce an FRS in our framework, whose results will serve as the real adversarial samples. We incorporate a pre-search and a filtering operation in the FRS. According to [23], the degeneration of underwater vision is caused by absorption, forward scattering, and backward scattering. The wavelength λ, water depth de, and object-to-camera distance di are related to the degeneration. We treat this problem using simplifying assumption, i.e., color distortion is produced through absorption, and haziness is produced from forward and backward scattering. Mathematically, l and q are used to denote original and degenerated signals. For wavelength λ, the absorption can be formulated with an exponential decay, i.e., $q_\lambda^{abs} = e^{-\eta \cdot di} l_\lambda$, where q_λ^{abs} is the absorbed signal, $e^{-\eta \cdot di}$ is the scalar absorption term that multiplies with each element in a matrix, and $\eta = \eta(\lambda, de)$ denotes absorption factor. Then, haziness can be expressed by convolution, i.e., $q_\lambda^{scatter} = h * q_\lambda^{abs}$, where $q^{scatter}$ represents scattered signal; $h = h(de, di)$ indicates a hazing convolution template related to forward and backward scattering. Ambient illumination sources are also merged into the signal transmission path during backward scattering, so the final degenerated signal $q_\lambda = q_\lambda^{scatter} + n$, where $n = n(de, di)$ is noise term. Thus, the degeneration model is formulated as follows:

$$q_\lambda = h * q_\lambda^{abs} + n. \tag{2.2}$$

The dehazing is thereby converted into a deconvolution task. In the next step, the convolution operation inspires us to apply a Fourier transform. Hence, the above analysis can be transferred to the Fourier domain:

$$Q_\lambda(u,v) = \mathcal{H}(u,v).*Q_\lambda^{abs}(u,v) + \mathcal{N}(u,v), \tag{2.3}$$

where the symbol ".*" denotes element-wise multiplication for a matrix. The turbulence model proposed by Hufnagel and Stanley [24] is used to formulate $\mathcal{H}$:

$$\mathcal{H}(u,v) = e^{-k(u^2+v^2)^{5/6}}, \tag{2.4}$$

where u and v are frequency variables, and k is associated with the intensity of a turbulent medium. Note that k wraps de and di, i.e., $k = k(de, di)$. To obtain an unambiguous image content, the Wiener filter is employed as follows:

$$\hat{Q}_\lambda^{abs}(u,v) = \left[\frac{\mathcal{H}^c(u,v)}{\left|\mathcal{H}(u,v)\right|^2 + R(u,v)} \right] Q_\lambda(u,v), \tag{2.5}$$

where $\mathcal{H}^c(u,v)$ denotes the conjugate matrix of $\mathcal{H}(u,v)$. Related to de and di, $R(u,v)$ is the noise to signal ratio that suppresses the effect of ambient illumination sources. It can be seen that the FRS generates $\hat{Q}_\lambda^{abs}(u,v)$ rather than the ideal restoration, and we use $\hat{Q}_\lambda^{abs}(u,v)$ as the real adversarial sample since it has been able to present key features. The FRS also requires information estimation of k, R, and they can be estimated by optimization approaches; e.g., [13] used artificial fish swarm algorithm for parameter search. Thus, the limits of applicability of the FRS are remarkable. That is, k and R are fragile with the change of underwater environments, and the search for them is computationally expensive. Inversely, the GAN-RS forgoes the need for any prior knowledge.

2.3.2 Architecture of the GAN-Based Restoration Scheme

As illustrated in Figure 2.2, the proposed architecture includes a generator G and a discriminative model D, and D contains an adversarial branch and a critic branch D_c.

The generator G based on a forward CNN is an encoder-decoder structure [17], which is composed of D_a residual blocks. By means of a 9-residual-block stack, the downsample-upsample model learns the essence of the input scene, and a synthesized version will emerge at the original resolution after the deconvolution operations.

We design the discriminator D in a multibranch manner including an adversarial branch and a critic branch. Using an image group (i.e., an underwater image concatenates a G's output or an FRS's output) as the input, the multibranch structure analyzes images from two aspects with forward CNNs, followed by the generations of an adversarial map and an underwater index map. The trunk of D is a one-layer convolution, and for the purpose of preserving image content, the real-or-fake discrimination is realized through the adversarial branch. On the other hand, the critic branch is carefully designed as a regression to discern whether an image

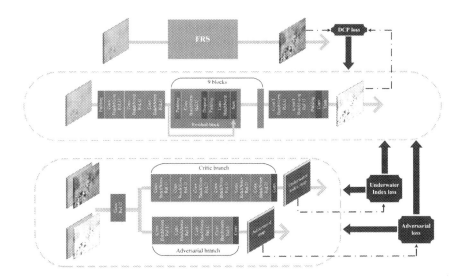

FIGURE 2.2 The architecture of the GAN-RS. From top to bottom: The FRS, generator, and discriminative model. The generator is a downsample–upsample framework with residual blocks. The multibranch discriminator is equipped with an adversarial branch and a critic branch. An adversarial loss, an underwater index loss, and a DCP loss are designed for training.

belongs to an underwater scene or not. That is, it evaluates the intensity of underwater property in an image, promoting the generator to produce images without underwater noise. These two branches are designed using a stack of Conv-BatchNorm-ReLU (CBR) units to concern image features. Inspired by the PatchGANs, we design both branches based on the idea of "patch", and the number of employed CBR units impacts the patch size (or receptive field). The effect of patch in adversarial training has been discussed by pix2pix [18], so we inherit its setting. As for the critic branch, learning underwater property needs more contextual information, so larger patch should have acquired better performance. However, the size of the underwater index map decreases with increasing patch size. Compared to small underwater index map, a large one is more effective for training owing to data augmentation. As a trade-off, we construct the adversarial branch with four CBR units, whereas the critic branch is built using six units. Finally, the resolution of the output underwater index map is 6×6, and the size of the receptive field is 286×286. In addition, the adversarial branch is constructed using two fewer CBR units, and thus the sizes become 30×30 and 70×70.

2.3.3 Objective for GAN-RS

2.3.3.1 Adversarial Loss

As the input condition, the original underwater image fed into G is denoted as x, and G tries to generate "real" sample y with noise z, i.e., $G(x,z) \rightarrow y$. The original conditional adversarial loss is a form of cross entropy [18]. However, Mao et al. stated that the cross entropy may lead to a problem with vanishing gradient during training, and they advocated the use of least squares generative adversarial networks (LSGANs) [25], whose loss function is the following least squares form:

$$\mathcal{L}_{lscGAN_D} = \mathbb{E}_{x,y \sim p_{data}(x,y)} \left[(D_a(x,y) - a)^2 \right]$$

$$+ \mathbb{E}_{x \sim p_{data}(x), z \sim p_z(z)} \left[(D_a(x, G(x,z)) - b)^2 \right] \quad (2.6)$$

$$\mathcal{L}_{lscGAN_G} = \mathbb{E}_{x \sim p_{data}(x), z \sim p_z(z)} \left[(D_a(x, G(x,z)) - a)^2 \right],$$

where $p_{data}(x,y)$, $p_z(z)$ represent x–y joint distribution and noise distribution, respectively. Hence, the LSGANs is employed for efficiency, and $a = 1$, $b = 0$ are the labels of the real or synthesized data, respectively.

2.3.3.2 DCP Loss

To promote the generator to not only fool the discriminator but also encourage an output close to the ground truth at the pixel level, an L1 loss between y and $G(x,z)$ is employed in pix2pix. Moreover, the effectiveness of this L1 loss has been verified by [18]. Because the FRS-processed samples are not final results, we do not expect a pixel-level similarity. Therefore, our method for the underwater image restoration task cannot employ this L1 loss, and we design a DCP loss based on the knowledge that there is a distinctive appearance between a hazy image and its clear version in a dark channel [14]. Here, we compute a dark channel for each pixel p and construct the DCP loss as follows:

$$y_{dark}(p) = \min_{\lambda \in \{R,G,B\}} y_\lambda(p) G(x,z)_{dark}(p)$$

$$= \min_{\lambda \in \{R,G,B\}} G(x,z)_\lambda(p) \quad (2.7)$$

$$\mathcal{L}_{DCP} = \mathbb{E}_{x,y \sim p_{data}(x,y), z \sim p_z(z)} \left\| y_{dark} - G(x,z)_{dark} \right\|.$$

2.3.3.3 Underwater Index Loss

If D works only using the adversarial branch, the networks cannot outperform FRS since the GAN would consider the real samples as the ideal outputs. Thus, to further improve the visual quality and promote G to generate underwater-noise-removed images, a novel loss function is proposed to train the critic branch, namely, the underwater index loss. According to the observation of massive amounts of data, we deem that there is a distinctive characteristic of underwater images in the Lab color space. Referring to Figure 2.3a, the Lab color space has a strong capability to indicate a color distribution, i.e., dark gray (right block) and light gray (left block) can be clearly differentiated on the a-axis, whereas black (bottom block) and white (top block) can be discerned on the b-axis. Moreover, as shown by the upper left pattern in Figure 2.3a, the a–b scatters of an underwater scene consistently gather far from the origin (shown with the upper left pattern), whereas those of an in-air image usually distribute sparsely with the origin as the center (shown with the central pattern). Thus, three distances, i.e., d_a, d_b, and d_o, can be used to formulate the possibility of an image having been taken underwater. Accordingly, the underwater index is given as

$$U = \frac{\sqrt{d_o}}{10a_l d_a d_b}, \tag{2.8}$$

where a_l denotes the average value of the L-channel, and the square root for d_o is employed for the purpose of amplifying a small distance.

Next, the underwater index loss is designed, which is learned using L2 sense function by the critic branch:

$$\mathcal{L}_{UD} = \mathbb{E}_{y \sim p_{data}(y)}\left[(D_c(y) - U(y))^2\right]$$
$$+ \mathbb{E}_{x \sim p_{data}(x), z \sim p_z(z)}\left[(D_c(G(x,z)) - U(G(x,z)))^2\right] \tag{2.9}$$
$$\mathcal{L}_{UG} = \mathbb{E}_{x \sim p_{data}(x), z \sim p_z(z)}\left[(D_c(G(x,z)))^2\right].$$

where $U(\cdot)$ computes the underwater index of an image. From (2.9), it can be seen that $\mathcal{L}_{UG}$ is trained toward 0 rather than real samples, so the real samples are not ideal outputs in our design, and the GAN-RS is able to perform better than the FRS.

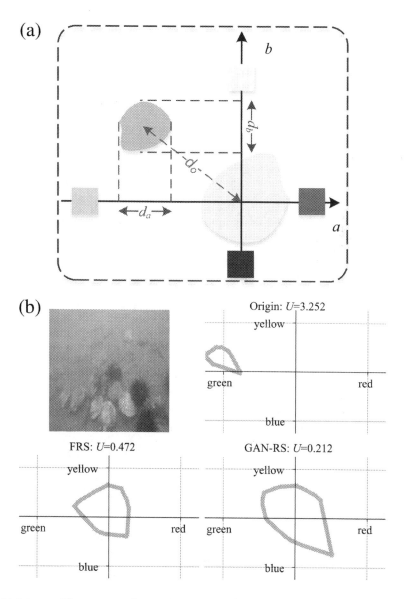

FIGURE 2.3   Illustration of an underwater index. (a) A diagram of the underwater index in Lab color space. The upper left pattern denotes the a-b distribution of an image, and d_o, d_a, d_b can be used to discriminate an underwater image and in-air image. (b) Typical experiment results. U is larger in terms of the original frame and is thus probably an underwater image, whereas the a-b scatters of a GAN-RS-processed frame is closer to the in-air distribution. Hard to be visualized, the scatter points are excessively dense, so we present their convex hulls.

2.3.3.4 Full Loss
The full objective for ACL is

$$
\begin{aligned}
\mathcal{L}_D &= \omega_{GAN}\mathcal{L}_{lscGAN_D} + \omega_U\mathcal{L}_{U_D}\mathcal{L}_G \\
&= \omega_{GAN}\mathcal{L}_{lscGAN_G} + \omega_U\mathcal{L}_{U_G} + \omega_{DCP}\mathcal{L}_{DCP},
\end{aligned}
\tag{2.10}
$$

where ω is the trade-off parameters, and the optimal models are formulated as $D^* = \arg_D \min \mathcal{L}_D, G^* = \arg_G \min \mathcal{L}_G$.

Despite two branches in D, our models can be trained following the canonical GAN paradigm. That is, G and D use their respective optimizers for back-propagation so that they can be trained individually and simultaneously. Unlike the traditional GAN, our discriminator generates two losses (i.e., an adversarial loss and an underwater index loss), and we add them up for back-propagation based on (2.10).

2.4 EXPERIMENTS AND DISCUSSION

2.4.1 Details of ACL

2.4.1.1 Basic Settings
By collecting underwater images on the seabed in China, a training set was established with 2,201 images, whereas the test set combined our data with public underwater images. Our training setting is according to the DCGAN [26]. The learning rate begins at 0.0002, and a linear decay is employed after 50 epochs. There are 65-epoch iterations in total. The Adam solver is employed as the optimizer [27]. Both the input and output resolutions are 512×512. Additionally, the selection of ω is important. For example, if ω_U is too large, G can rapidly generate images with lower underwater index, but, simultaneously, D will quickly distinguish real and fake samples. As a result, $\mathcal{L}_{GAN}$ loses its effect, and G could bring about artifacts in the output images. If ω_U is too small, $\mathcal{L}_U$ cannot impact the training process. Experimentally, $\omega_{GAN} = 1, \omega_U = 10, \omega_{DCP} = 30$ are selected based on the model performance, and these training parameters can assure the stable training of GAN. Note that ω is only used for training, which would not appear in the test phase. Thus, for real-world applications, the GAN-RS does not depend on any experimental or empirical parameter.

2.4.1.2 Multistage Loss Strategy
We develop a multistage loss strategy for effective training, i.e., $\mathcal{L}_G = \omega_{GAN}\mathcal{L}_{lscGAN_G} + \omega_{DCP}\mathcal{L}_{DCP}$ at the beginning of training. Then, $\omega_U\mathcal{L}_{U_G}$

will be added to $\mathcal{L}_G$ at a specific timestamp (i.e., the 30th epoch in this chapter). The necessity of the multistage loss strategy is twofold: (1) The critic branch randomly predicts underwater index map at the beginning of training, so the $\mathcal{L}_{U_G}$ is worthless until D_c has been optimized; (2) this operation eliminates the early impact of $\mathcal{L}_{U_G}$, so that D_a and G can achieve a dynamic equilibrium as soon as possible for stable training.

The loss curves are illustrated in Figure 2.4b. A stair in $\mathcal{L}_{U_G}$ is evident when the critic branch goes into effect. Moreover, in terms of adversarial loss, it can be seen that G and D achieve dynamic equilibrium early in the training (i.e., $\mathcal{L}_{lscGAN_G} \approx 0.30$, $\mathcal{L}_{lscGAN_G} \approx 0.22$), whereas $\mathcal{L}_{GAN}$ deviates from the balance points when $\mathcal{L}_{U_G}$ is applied. That is, the generated image is deemed to probably be synthesized, whereas D is more certain about its judgment. Gradually, a new dynamic equilibrium will be obtained at another pair of balance points (i.e., $\mathcal{L}_{lscGAN_G} \approx 0.40$, $\mathcal{L}_{lscGAN_G} \approx 0.19$).

2.4.2 Compared Methods

The proposed methods are compared with the GW [23], CLAHE [28], probability-based method (PB) [29], RBLA [7], pix2pix [18], CycleGAN [21], and the dehazing (DM) or contrast enhancement method (CM) in [6]. All the abovementioned methods are implemented with open source codes. As for pix2pix [18] and CycleGAN [21], they are under PyTorch framework, and we adjust training parameters to train our dataset more effectively. On the contrary, other methods are based on Matlab, and we maintain the original parameters in their papers. It should be remarked that the comparison between the pix2pix and GAN-RS can unveil the effectiveness of critic branch and underwater index loss.

2.4.3 Runtime Performance

2.4.3.1 Running Environment

The GAN-RS is implemented under the PyTorch framework. Experiments are carried out on a workstation with an Intel 2.20 GHz Xeon(R) E5-2630 CPU, an NVIDIA TITAN-Xp GPU, and 64 GB RAM.

2.4.3.2 Time Efficiency

All the runtime data are tested using 512×512 images. Most approaches are based on Matlab, but our FRS is a C++ project while the GAN-RS is implemented under a GPU-based framework. Thus, the speed of CLAHE is described in the form of *Matlab speed/C++speed*, whereas the FRS speed

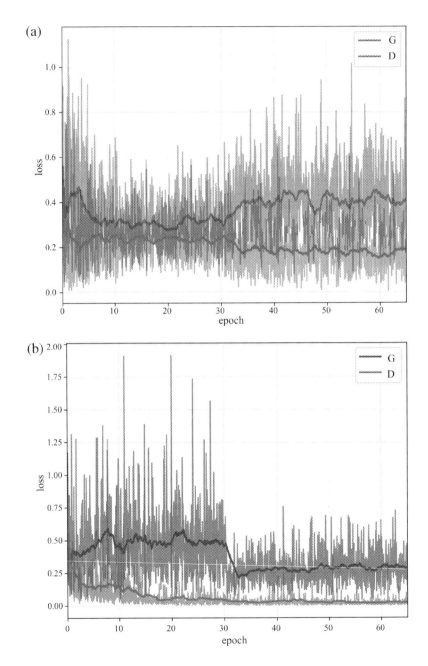

FIGURE 2.4 Illustration of multistage training loss in ACL: (a) Adversarial training process and its losses $\mathcal{L}_{bcGAN}$; (b) critical learning process and its lossed $\mathcal{L}_v$. The stair in $\mathcal{L}_{v,G}$ indicates that the critic branch goes into effect, and the $\mathcal{L}_{bcGAN}$ achieves dynamic equilibrium twice.

TABLE 2.1   FPS list by the proposed methods and several contemporary approaches

Method	FPS	Method	FPS
GW [24]	18.21	RBLA [7]	0.02
PB [26]	1.45	CLAHE [25]	21.27/84.03
DM [6]	0.43	FRS	38.91/118.56
CM [6]	0.32	GAN-RS	133.77

is presented as *CPU speed/GPU speed*. Sharing the runtime performance, the pix2pix and CycleGAN use the same G as the GAN-RS, so they are not listed. As shown in Table 2.1, the processing speed for FRS is 118.56 FPS. Moreover, far superior to the existing restoration methods, the GAN-RS reaches 133.77 FPS.

2.4.4 Restoration Results

2.4.4.1 Visualization of Underwater Index

As an illustrative example, the underwater index is delineated graphically in Figure 2.3b. The demonstrated image is a typical underwater environment, which is quite hazy and color distorted. The upper right corner of Figure 2.3b shows original color distribution in the a-b plane. As can be seen, the color distortion is reflected in the distance between the distribution center and the origin, i.e., d_o is large for terrible color distortion. On the other hand, the haziness is related to the concentration of the distribution. Briefly, $d_a d_b$ approaches to 0 owing to the haziness or lower contrast, and thus $U \to 0$ is the ideal condition. Although U is not involved in the optimization, the FRS performs well to enhance the underwater index (see the lower left corner of Figure 2.3b). Further, the GAN-RS uses a critic branch to decrease the underwater index loss, so a more considerable U can be achieved with less bias and greater dispersion in the a-b plane (see the lower-right corner of Figure 2.3b). Therefore, the underwater index has the capability of describing the underwater property intensity in an image.

2.4.4.2 Comparison on Restoration Quality

The comparison, shown in Figure 2.5, verifies the qualitative superiority of the proposed GAN-RS. Compared with several prior and contemporary methods, our method achieves a clearer vision, more balanced color, and stretched contrast. As can be seen, some approaches see limited effect as for restoration quality, e.g., the GW only achieves a white balance;

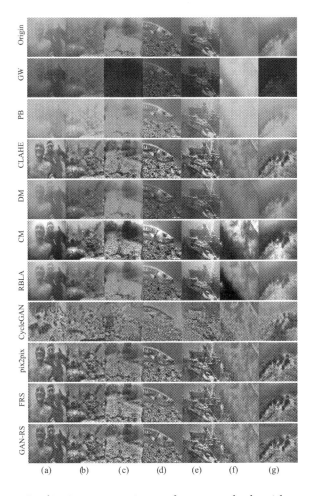

FIGURE 2.5 Qualitative comparison of our method with contemporary approaches in terms of restoration quality. Images in each column are restored by the denoted method. (a–c) are collected by [9]; (d) is collected by [10]; (e–f) is collected by [5]; (g) is collected by [8].

the CLAHE has an insignificant effect on the color correction; and the brightness advancement introduced by PB comes with an aggravation of the color distortion. Meanwhile, due to supervised adversarial training, our method results in little damage to the original image content. On the contrary, the CycleGAN cannot maintain the semantic content owing to lack of effective supervision, whereas the CM cannot preserve the objective color of an image in certain cases (see Figure 2.5). The RBLA performs well, but there is a drawback that the parameter adjustment is complex and empirical. For instance, the RBLA restores Figure 2.5a at a resolution of 404×303 in this chapter [7]. In this chapter, however, a 512×512 version is applied instead, and its performance is restricted with original parameters.

The numerical comparison is shown in Table 2.2. There is no in-air ground truth for comparison, and therefore, some no-reference quality assessment tools are employed, including the underwater index, Laplace gradient, entropy, underwater color image quality evaluation (UCIQE) metric [30], and underwater image quality measure (UICM, UISM, UIConM, UIQM) [31]. The underwater index proposed in this chapter can be treated as an underwater property intensity in an image. The Laplace gradient reflects haze degree, whereas the entropy denotes richness of image information. The UIQM, composed of UICM, UISM, and UIConM, represents a comprehensive quality of a restored underwater image, and its sub-indexes are the pros and cons of the color, sharpness, and contrast. Similarly, the UCIQE quantifies image quality through the chrominance, average saturation, and luminance contrast. Note that CycleGAN and CM are not involved in owing to the abovementioned drawback. As shown in Table 2.2, results include two typical underwater environments (i.e., (a) and (k)) and the average among tested images. Some methods work well from a particular perspective; e.g., the GW is effective against color distortion, and the RBLA generated the best production for UCIQE. In terms of underwater index, it is interesting to note that the pix2pix achieves a performance similar to but not better than its ground truth (FRS), whereas the GAN-RS achieves a significant improvement in d_o, $d_a d_b$, and U credited to the critic branch. As for UIQM, the FRS generates UICM-optimal outputs, and the GAN-RS is better with regard to UISM, UIConM, and UIQM. Therefore, it can be concluded that the comprehensive performance of the proposed GAN-RS is better in terms of the restoration quality.

2.4.4.3 Feature-Extraction Tests

In this subsection, some feature-extraction algorithms, including SIFT [32], Harris [33], and Canny [34] are employed to test the application of the GAN-RS from the perspectives of fundamental features and object detection. As shown in Figure 2.6, few key points can be obtained by SIFT in the original frame, and a correct match seldom occurs. There are limited improvements brought by most compared methods. On the contrary,

TABLE 2.2 Quantitative comparison using no-reference quality assessment. (a), (g) are with regard to Figure 2.5

Label	Method	d_o	$d_a d_b$	U	Laplace	Entropy	UCIQE	UICM	UISM	UIConM	UIQM
(a)	Origin	0.79	0.04	3.60	3.11	6.93	0.42	-0.63	3.19	0.15	1.45
	GW	0.51	0.25	0.53	2.03	6.54	0.45	2.83	1.75	0.12	1.02
	PB	0.84	0.03	3.76	2.12	6.47	0.36	-2.01	1.83	0.10	0.85
	CLAHE	0.58	0.11	0.97	8.21	7.22	0.49	1.11	1.83	0.17	4.13
	DM	0.79	0.04	3.95	2.72	6.91	0.45	0.06	3.26	0.13	1.43
	RBLA	0.63	0.13	1.02	3.93	7.51	0.57	2.21	5.75	0.15	2.30
	pix2pix	0.35	0.15	0.47	16.50	7.23	0.49	2.20	3.87	0.21	4.90
	FRS	0.36	0.14	0.51	23.10	7.24	0.51	2.51	14.42	0.21	5.07
	GAN-RS	0.22	0.2	0.25	17.96	7.15	0.50	2.63	4.20	0.21	5.01
(g)	Origin	0.74	0.04	4.28	6.22	6.49	0.42	0.14	6.47	0.17	2.51
	GW	0.53	0.13	2.65	3.26	5.28	0.49	4.43	3.55	0.14	1.68
	PB	0.83	0.05	2.64	4.28	6.13	0.36	-0.52	5.92	0.12	2.17
	CLAHE	0.49	0.17	0.59	19.70	7.01	0.53	3.83	11.63	0.17	4.15
	DM	0.72	0.05	3.45	6.27	6.43	0.42	0.06	6.03	0.14	2.27
	RBLA	0.53	0.12	0.92	10.65	7.03	0.55	7.03	8.86	0.16	3.38
	pix2pix	0.35	0.23	0.36	24.76	7.09	0.56	4.08	12.76	0.20	4.60
	FRS	0.28	0.19	0.34	25.12	7.13	0.57	4.02	13.42	0.20	4.79
	GAN-RS	0.3	0.27	0.25	28.80	7.06	0.57	4.19	13.57	0.21	4.86
Average	Origin	0.66	0.05	2.81	4.77	6.46	0.44	0.26	5.56	0.15	2.20
	GW [24]	0.51	0.16	1.28	3.52	6.03	0.46	3.03	4.24	0.13	1.80
	PB [26]	0.73	0.06	2.51	3.67	6.11	0.40	-0.46	5.02	0.11	1.87
	CLAHE [25]	0.48	0.14	0.69	**12.28**	7.11	0.52	2.35	11.33	0.16	3.98
	DM [6]	0.64	0.07	2.32	4.69	6.44	0.46	0.99	5.52	0.14	2.14
	RBLA [7]	0.51	0.12	0.93	7.10	7.11	**0.56**	3.93	8.30	0.15	3.09
	pix2pix [18]	0.25	0.18	0.32	20.51	7.18	0.53	2.74	13.58	**0.20**	4.79
	FRS	0.26	0.19	0.30	23.85	**7.26**	0.55	**3.14**	13.75	**0.20**	4.85
	GAN-RS	**0.19**	**0.21**	**0.20**	22.95	7.19	0.54	2.83	**13.87**	**0.20**	**4.88**

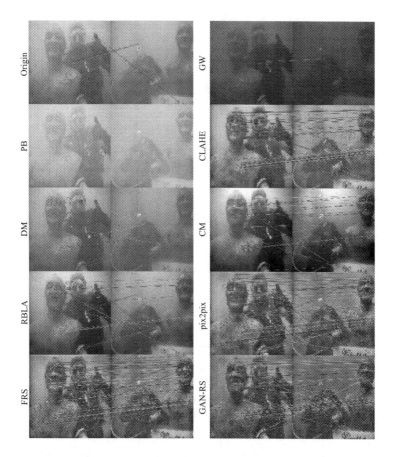

FIGURE 2.6   Feature-extraction tests: SIFT match.

assisted by the GAN-RS, salient features are extracted, and a multitude of accurate matches appear.

The numerical comparison is shown in Table 2.3, where "SIFT, Harris" are the number of SIFT key points and Harris corners; "Canny" computes the pixel-level edge ratio in an image. By comparison, the SIFT and Harris perform better when combined with GAN-RS, whereas the output of FRS covers more edges. Therefore, it is verified that the proposed GAN-RS contributes to the extraction of fundamental features of underwater images.

2.4.5 Visualization of Discriminator

The evaluation of restoration quality can validate the effectiveness of the generator, but the performance of G comes from the discriminator and

TABLE 2.3   Numerical comparison results

Label	Figure	SIFT	Harris	Canny
(a)	Origin	61	0	0.00
	GW	20	0	0.00
	PB	20	0	0.00
	CLAHE	628	278	0.04
	DM	94	12	0.00
	CM	373	227	0.03
	RBLA	256	121	0.20
	pix2pix	1,732	1,522	0.11
	FRS	1,154	1,652	0.18
	GAN-RS	1,804	1,633	0.14
Average	Origin	626.73	380.36	0.03
	GW	536.09	427.27	0.03
	PB	784.09	530.82	0.03
	CLAHE	2,372.18	1,796.82	0.13
	DM	1,036.09	497.36	0.04
	CM	2,143.91	1,877.73	0.15
	RBLA	1,361.91	1,080.55	0.08
	pix2pix	2,508.55	2,289.82	0.15
	FRS	2,288.45	2,537.91	**0.19**
	GAN-RS	**2,632.27**	**2,556.00**	0.17

ACL. As shown in Figure 2.7, we visualize adversarial map and underwater index map to unveil the effect of D. The adversarial branch is able to roughly distinguish FRS and GAN-RS outputs. As a result, adversarial map of FRS has higher discrimination saliency. In spite of this, the content and structure in an image are preserved in adversarial learning. In the meantime, the critic branch is able to remove underwater noise, so the underwater index map of GAN-RS has lower saliency than that of FRS. Therefore, both adversarial learning and critical learning are significant in ACL so that the performance of generator can be further improved.

2.4.6 Discussion

Quality advancement of underwater robotic vision is essential to underwater visual-based operation and navigation. We creatively treat the restoration task as an image-to-image translation to enhance the real-time capacity and adaptability. We use the FRS to supervise the adversarial

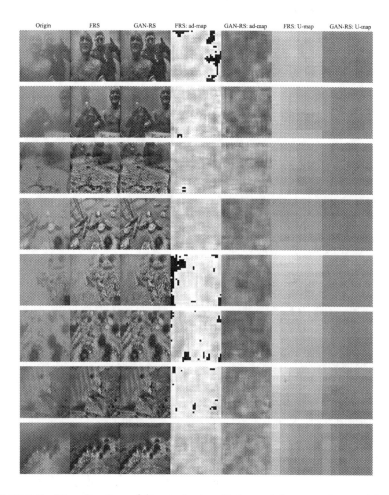

FIGURE 2.7   Visualization of discriminator. "Ad-map" denotes adversarial map, while "U-map" is underwater index map.

branch, but compared to the FRS, the GAN-RS has merits in processing speed, restoration quality, and adaptability. Moreover, there exists room for improvement: (1) Attention mechanism [35] for selective adversarial training could be beneficial, and (2) the GAN-RS could perform better if real samples come from multiple traditional approaches.

The limitations of GAN-RS are twofold. On the one hand, collecting underwater samples is a costly work. On the other hand, the training parameters need to be carefully set or adjusted. The generative model could bring about artifacts in the output images if trained using an improper setting.

2.5 CONCLUDING REMARKS

In this chapter, we aim at adaptively restoring underwater images in real time and propose a GAN-RS. Differing from the existing methods, the GAN-RS restores underwater visual quality using a single-shot network for higher computational efficiency and greater restoration quality. A multibranch discriminator is designed including an adversarial branch and a critic branch to promote the generator to simultaneously preserve image content and remove underwater noise. In addition, an underwater index loss is investigated based on the underwater properties, and a DCP loss as well as a multistage loss strategy is developed for training. As a result, the proposed GAN-RS adaptively restores underwater scenes at a high frame rate. Moreover, both qualitative and quantitative comparisons on restoration quality and feature extraction are conducted, and the GAN-RS achieves a comprehensively superior performance in terms of visual quality and feature restoration. Finally, our proposed approach has been employed in a practical application on the seabed for object grasping to achieve encouraging results.

REFERENCES

1. M.-C. Chuang, J.-N. Hwang, K. Williams, and R. Towler, "Tracking live fish from low-contrast and low-frame-rate stereo videos," *IEEE Trans. Circuits Syst. Video Technol.*, vol. 25, no. 1, pp. 167–179, 2015.
2. X. Chen, J. Yu, and Z. Wu, "Temporally identity-aware SSD with attentional LSTM," *IEEE Trans. Cybern.*, vol. 50, no. 6, pp. 2674–2686, 2020.
3. W. Liu, D. Anguelov, D. Erhan, C. Szegedy, S. Reed, C.-Y. Fu, and A.-C. Berg, "SSD: Single shot multibox detector," in *Proceedings of the European Conference on Computer Vision*, Amsterdam, Netherlands, Oct. 2016, pp. 21–37.
4. Y.-Y. Schechner and N. Karpel, "Clear underwater vision," in *Proceedings of the IEEE Conference on Computer Vision and Pattern Recognition*, Washington, USA, Jun. 2004, pp. I-536–I-543.
5. J.-Y. Chiang and Y. Chen, "Underwater image enhancement by wavelength compensation and dehazing," *IEEE Trans. Image Process.*, vol. 21, no. 4, pp. 1756–1769, 2012.
6. C. Li, J. Guo, R. Cong, Y. Pang, and B. Wang, "Underwater image enhancement by dehazing with minimum information loss and histogram distribution prior," *IEEE Trans. Image Process.*, vol. 25, no. 12, pp. 5664–5677, 2016.
7. Y.-T. Peng and P. C. Cosman, "Underwater image restoration based on image blurriness and light absorption," *IEEE Trans. Image Process.*, vol. 26, no. 4, pp. 1579–1594, 2017.

8. S. Emberton, L. Chittka, and A. Cavallaro, "Hierarchical rank-based veiling light estimation for underwater dehazing," in *Proceedings of the British Machine Vision Conference*, Swansea, UK, Sep. 2015, pp. 125.1–125.12.

9. C. Ancuti, C. O. Ancuti, T. Haber, and P. Bekaert, "Enhancing underwater images and videos by fusion," in *Proceedings of the IEEE Conference on Computer Vision and Pattern Recognition*, Providence, Rhode Island, Jun. 2012, pp. 81–88.

10. Galdran, D. Pardo, A. Picon, and A. Alvarez-Gila, "Automatic red-channel underwater image restoration," *J. Visual Commun. Image Represent.*, vol. 26, pp. 132–145, 2015.

11. Goodfellow, J. Pouget-Abadie, M. Mirza, B. Xu, D. Warde-Farley, S. Ozair, and Y. Bengio, "Generative adversarial nets," in *Proceedings of the Advances in Neural Information Processing Systems*, Montreal, Canada, Dec. 2014, pp. 2672–2680.

12. Johnson, A. Alahi, and L. Fei-Fei, "Perceptual losses for real-time style transfer and super-resolution," in *Proceedings of the European Conference on Computer Vision*, Amsterdam, Netherlands, Oct. 2016, pp. 694–711.

13. X. Chen, Z. Wu, J. Yu, and L. Wen, "A real-time and unsupervised advancement scheme for underwater machine vision," in *Proceedings of the IEEE International Conference on CYBER Technology in Automation, Control, and Intelligent Systems*, Hawaii, USA, Aug. 2017, pp. 271–276.

14. He, J. Sun, and X. Tang, "Single image haze removal using dark channel prior," *IEEE Trans. Pattern Anal. Mach. Intell.*, vol. 33, no. 12, pp. 2341–2353, 2011.

15. Y.-S. Shin, Y. Cho, G. Pandey, and A. Kim, "Estimation of ambient light and transmission map with common convolutional architecture," in *Proc. IEEE/MTS OCEANS*, Monterey, USA, Sep. 2016, pp. 1–7.

16. J.-Y. Zhu, P. Krahenbuhl, E. Shechtman, and A.-A. Efros, "Generative visual manipulation on the natural image manifold," in *Proceedings of the European Conference on Computer Vision*, Amsterdam, Netherlands, Oct. 2016, pp. 597–613.

17. C. Ledig, L. Theis, F. Huszar, J. Caballero, A. Cunningham, A. Acosta, and W. Shi, "Photo-realistic single image super-resolution using a generative adversarial network," in *Proceedings of the IEEE Conference on Computer Vision and Pattern Recognition*, Honolulu, USA, Jun. 2017, pp. 4681–4690.

18. P. Isola, J.-Y. Zhu, T. Zhou, and A.-A. Efros, "Image-to-image translation with conditional adversarial networks," in *Proceedings of the IEEE Conference on Computer Vision and Pattern Recognition*, Honolulu, USA, Jun. 2017, pp. 1125–1134.

19. Mirza and S. Osindero, "Conditional generative adversarial nets," https://arxiv.org/abs/1411.1784, 2014.

20. H. Dong, P. Neekhara, C. Wu, and Y. Guo, "Unsupervised image-to-image translation with generative adversarial networks," https://arxiv.org/abs/1701.02676, 2017.

21. J.-Y. Zhu, T. Park, P. Isola, and A.-A. Efros, "Unpaired image-to-image translation using cycle-consistent adversarial networks," in *Proceedings of the IEEE International Conference on Computer Vision*, Venice, Italy, Oct. 2017, pp. 2223–2232.
22. M.-Y. Liu, T. Breuel, and J. Kautz, "Unsupervised image-to-image translation networks," in *Proceedings of the Advances in Neural Information Processing Systems*, Long Beach, USA, Dec. 2017, pp. 700–708.
23. E. Provenzi, C. Gatta, M. Fierro, and A. Rizzi, "A spatially variant white-patch and gray-world method for color image enhancement driven by local contrast," *IEEE Trans. Pattern Anal. Mach. Intell.*, vol. 30, no. 10, pp. 1757–1770, 2008.
24. R. Hufnagel and N. Stanley, "Modulation transfer function associated with image transmission through turbulent media," *J. Opt. Soc. Am.*, vol. 54, no. 1, pp. 52–60, 1964.
25. X. Mao, Q. Li, H. Xie, R.-Y. Lau, Z. Wang, and S.-P. Smolley, "Least squares generative adversarial networks," in *Proceedings of the IEEE International Conference on Computer Vision*, Venice, Italy, Oct. 2017 pp. 2794–2802.
26. Radford, L. Metz, and S. Chintala, "Unsupervised representation learning with deep convolutional generative adversarial networks," *arXiv:1511.06434*, 2015.
27. D. Kingma and J. Ba, "Adam: A method for stochastic optimization," https://arxiv.org/abs/1412.6980, 2014.
28. K. Zuiderveld, "Contrast limited adaptive histogram equalization," in *Graphics gems IV*, Elsevier, London, pp. 474–485, 1994.
29. X. Fu, Y. Liao, D. Zeng, Y. Huang, X.-P. Zhang, and X. Ding, "A probabilistic method for image enhancement with simultaneous illumination and reflectance estimation," *IEEE Trans. Image Process.*, vol. 24, no. 12, pp. 4965–4977, 2015.
30. Yang and A. Sowmya, "An underwater color image quality evaluation metric," *IEEE Trans. Image Process.*, vol. 24, no. 12, pp. 6062–6071, 2015.
31. K. Panetta, C. Gao, and S. Agaian, "Human-visual-system-inspired underwater image quality measures," *IEEE J. Ocean. Eng.*, vol. 41, no. 3, pp. 541–51, 2015.
32. D.-G. Lowe, "Distinctive image features from scale-invariant keypoints," *Int. J. Comput. Vis.*, vol. 60, no. 2, pp. 91–110, 2004.
33. C. Harris and M. Stephens, "A combined corner and edge detector," in *Proceedings of the Alvey Vision Conference*, Manchester, UK, pp. 147–151, 1988.
34. J, Canny, "A computational approach to edge detection," *IEEE Trans. Pattern Anal. Mach. Intell.*, vol. 6, pp. 679–698, 1986.
35. H. Zhang, I. Goodfellow, D. Metaxas, and A. Odena, "Self-attention generative adversarial networks," *arXiv:1805.08318*, 2018. PMLR, vol. 97, pp. 7354–7363, 2019.

CHAPTER 3

A NSGA-II-Based Calibration for Underwater Binocular Vision Measurement

3.1 INTRODUCTION

Underwater image processing technology has attracted great attention in recent years. Compared to the acoustic sensors which are extensively applied in the underwater exploration and measurement [1,2], the underwater vision is more appropriate for the short distance operation, for its flexibility, portability, and low cost.

In practice, divers generally use the rulers to measure the dimension of various underwater objects, which is a labor-consuming, dangerous, and disruptive task. However, the well-known binocular vision system, as a replacement of the manual measurement, is a kind of non-contact and non-destructive sensor for measurement including position, length, and angle of objects, whose feasibility has been proved on the terrene. If the robotic fishes, autonomous underwater vehicles, remotely operated vehicles (ROV), and even divers can be equipped with the binocular vision system to measure underwater objects accurately, then the task for the underwater measurement will be more convenient and efficient.

The underwater binocular vision measurement system (UBVMS) will be applied in the field, including marine archaeology, marine exploration, and marine fishery. Supported by the National Natural Science Foundation of China and cooperated with Beihang University, the UBVMS is installed on an ROV to grasp seafood such as trepang, sea urchin, and scallop in marine ranching.

The calibrations for various camera systems in the air are fully investigated [3–6]. With the difference from vision system in air, the underwater vision system is confronted with great challenges. Employing the refractive model of cameras instead of the classical pinhole model is one of the most prominent challenges. The electric parts of cameras are normally protected by a waterproof house, which is usually made of refractive materials. Therefore, the light from the object to the optic center of the camera will pass through three different refractive index media, i.e., water, glass or acrylic, and air. Refraction will happen on two surfaces, like a water-to-glass and a glass-to-air surface. As a result, many widely used algorithms of binocular vision in the air are not applicative underwater anymore, which leads to inaccurate underwater measurement. Our calibration algorithm is proposed to solve this problem.

On the assumption of parallel refracting surfaces, we develop a mathematical model of the refractive camera to describe the light path through multiple media based on Snell's law. According to the mathematical model, there is a nonlinear relationship between the ray of light, the normal of refractive surfaces $\mathbf{n}_\pi$, the vertical distance from the optical center to the glass-to-air surface d_0, and the thickness of the ith media d_i ($i > 0$). We define $\mathbf{n}_\pi$, d_0, and d_i as the housing parameters. Then, we propose an akin triangulation method to compute the 3D coordinates of underwater objects, when the corresponding points on the image plane are known. Derived from the akin triangulation, a refractive surface constraint is employed to restrict the search space of the undermentioned optimization process. To avoid a complex mathematical analysis of solving the nonlinear refractive geometrical relationship equation, a flexible calibration algorithm with a novel usage of the checkerboard is put forward to obtain the housing parameters. This calibration process can be regarded as a multi-objective nonlinear optimization with nonlinear constraints. The NSGA-II, a reliable, intelligent optimization method, is suitable to solve this optimization problem. Finally, experimental results on underwater calibration and measurement demonstrate the effectiveness of the proposed calibration algorithm.

The primary contributions of this chapter are threefold. Firstly, we integrate the refractive model of multiple media without loss of generality, employ the underwater akin triangulation to obtain target positions, and provide a necessary refractive surface constraint to restrict the range of housing parameters. Secondly, a novel usage of checkerboard based on the relative position relationship of corners for underwater calibration is put forward to convert the process of underwater camera calibration to a multi-objective nonlinear optimization with nonlinear constraints, avoiding solving equation of the complex nonlinear refractive geometrical relationship. Thirdly, comprehensive experiments designed to measure the position of a checkerboard and the size of rulers in different postures demonstrate the feasibility and robustness of this calibration algorithm.

The rest of the chapter is organized as follows. Section 3.2 includes a brief overview of related works. In Section 3.3, the mathematical model of the multiple media refractive geometry is established. We propose an underwater binocular measurement method named akin triangulation in Section 3.4. The calibration algorithm for UBVMS based on NSGA-II is detailed in Section 3.5. Section 3.6 deals with experiments and analyses. Finally, the conclusion and future work are given in Section 3.7.

3.2 RELATED WORK

To achieve a good calibration for UBVMS, modeling nonlinear refractive relationship of underwater cameras is quite essential. Some comprehensive overviews on establishing camera models in underwater imaging and underwater calibration algorithms can be found in [7–10]. Shortis et al. offered an overview about the underwater calibration techniques, in which the underwater calibration methods are legitimately classified in three categories: The absorption, the geometric correction, and perspective center shift or virtual projection center approach [10]. Whereas the perspective center shift or the virtual projection center approach is a variation from geometric correction, for both of them are based on refractive photogrammetric models. Therefore, the main underwater calibration approaches can be divided into two categories substantially. Next, we give a brief overview of absorption calibration methods and geometric correction methods hereinafter.

While a lot of studies are conducted on underwater vision, by far, most of them do not establish refractive camera models explicitly [10], still based on the pinhole model as follows:

$$
\begin{pmatrix} x \\ y \\ 1 \end{pmatrix} = \begin{pmatrix} f_x & 0 & u_o & 0 \\ 0 & f_y & v_o & 0 \\ 0 & 0 & 1 & 0 \end{pmatrix} [R\,|\,T] \begin{pmatrix} X_w \\ Y_w \\ Z_w \\ 1 \end{pmatrix}.
\tag{3.1}
$$

Some of the above studies ignore the refractive effects completely [11–14], which do not belong to either of aforementioned two categories, and the experiments in this chapter present the quietly weak performance of them, whereas the absorption methods depend on absorption by the physical camera calibration parameters to correct the refraction effects. With the hypothesis that the camera optical axis is approximately per-pendicular to a plane or dome camera port, the primary effect of refrac-tion through the glass-to-air and water-to-glass interfaces will be radially symmetric around the optical center [15]. This primary effect of refraction can be absorbed by the radial lens distortion component of the calibra-tion parameters. Taking the refraction into consideration, dealing with its influence by improving the nonlinear distortion terms, is an implement of typical absorption methods [16,17]. Menna et al. utilized a network of digital still camera images to characterize the shape of a semi-submerged ship precisely, which is an example of absorption methods [18]. However, Treibitz et al. demonstrated that errors will always exist in underwater calibration by using a pinhole model [19]. The difference between the pin-hole model and the refractive model of cameras is shown in Figure 3.1. On account of the refraction, objects seem to be closer to the observer and be a little larger than they actually are. In this situation, the traditional tri-angulation based on the linear relationship of the multiple view geometry cannot derive the accurate 3D position information of underwater objects [20,21]. Besides, the main disadvantage of the absorption methods for the refractive effects was concluded in [10] that the assumption of a single projection center for the camera was unsuitable when the measurements were taken outside the range for the calibration process.

Recently, an increasing number of studies pay more attention to set-ting up and correcting refractive camera models since it can improve underwater measurement results. Geometric correction methods, with a

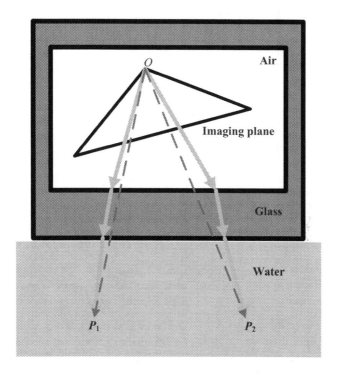

FIGURE 3.1 The difference between the pinhole model and the refractive model. O is the optical center of the camera. P_1 and P_2 are two objects underwater. The dotted lines represent the light path based on the pinhole model, while the solid lines denote the actual light path with two refractions on the surfaces. The two kinds of light rays with different models from objects to the optical center form different images on the image plane.

more complex process of refractive analysis, become the alternative to the simple approach of absorption. Li et al. proposed a two-phase calibration method based on quantitative photogrammetric analysis of underwater imagery, who were one of the pioneers for geometric correction methods [22]. Under the premise of refractions by flat surfaces, Sedlazeck et al. presented a flexible calibration method to obtain the housing parameters for underwater stereo rigs [23]. This method can account for two refractions without any calibration object by assuming that the glass thickness of the housing is known. Agrawal et al. proposed a special calibration technique based on the axial camera model to approximate exact model parameters of the refraction, such as the thickness of the refractive surface and its refractive indices [24]. In practice, the refractive indices are given, so the optimization process can be simplified. Subsequently, Jordt-Sedlazeck and Koch presented a new analysis-by-synthesis approach, coupled with an evolutionary algorithm for optimization, which allows calibrating the parameters of a light propagation model for the local water body [23]. Similar to the axis camera model, Chen et al. proposed a calibration method for an underwater stereo camera system but cannot avoid the complex mathematical analysis of the refractive relationship [25]. As a variation on the geometric correction, the perspective center shift and virtual projection center approaches were proposed by researchers. Telem et al. proposed a perspective center shift method to develop a solution for a planar housing port [26]. Dolereit et al. utilized virtual object points to calibrate underwater cameras, providing a novel way for researchers [27]. The time-of-flight correction based on Kinect Fusion between RGB and NIR was provided for 3D reconstruction in one of the recent researches, which fell within the scope of geometric correction methods [28]. In conclusion, the central purpose of geometric correction methods and its variations is to seek for an effective solution of refractive parameters [10]. The advantage of these techniques is that the correction of the refractive effects is exact and without approximations, while the disadvantages are requirements for two-phase calibrations and complex refractive geometrical analysis, and even depending on NIR. In this case, if two-phase calibrations and the geometrical analysis are necessary, a feasible calibration method with a simplified geometrical correction process and a convenient calibration tool are in need.

Inspired by geometrical correction methods [22–27,29], we integrate a complete refractive camera model with flat refractive surfaces and simplify the process of calibration by a multi-objective optimization to avoid solving the equation of the nonlinear refractive geometrical relationship. Compared to the method ignoring the refraction [11–14] and the method regarding the refraction as distortion terms, i.e., the original approach of absorption [16,17], as will be demonstrated later, our proposed calibration method for underwater cameras can improve the accuracy of the UBVMS.

3.3 REFRACTIVE CAMERA MODEL

In practice, the ray from objects to the optical center usually transmits through two refractive surfaces: The water-to-glass surface and the glass-to-air surface. To integrate the complete refractive model without loss of generality, we assume that the normal of the image plane is random, that there are N refractions between objects and the optical center, and that all the refractive surfaces are parallel to each other. This refractive model is primarily based on Snell's law. Note that the experimental setup of this chapter is to verify the feasibility of the proposed calibration method for UBVMS, because of which the model used in the experiment has only three specific media, i.e., air, glass, and water.

Firstly, the notations are described in Figure 3.2a. There are N refractive surfaces, and $\mu_i, i \in [0, N]$ denotes the refractive index of each medium.

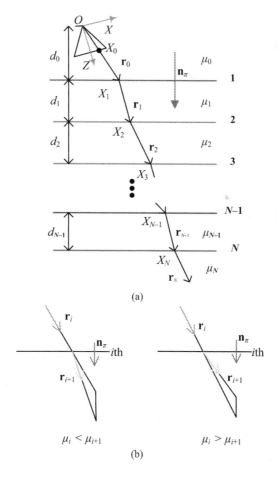

FIGURE 3.2 Underwater camera model considering refraction. (a) A general refractive camera model with N refractions. (b) Two geometrical relationships between the incident ray and the refracted ray in the condition of different comparisons of the refractive index.

O represents the optical center, on which the camera coordinate system is established. The directions of rays in each medium are indicated as $r_i, i \in [0, N]$, which can be regarded as a unit vector in the camera coordinate system. Meanwhile, the rays intersect each refractive surface at points $X_i, i \in [1, N]$. Let $d_i, i \in [1, N-1]$ be the thickness of each medium, but d_0 is the vertical distance from the optical center to the first refractive surface. The unit normal vector of the series of parallel interfaces is n_π, which is not parallel to the axis Z of the camera coordinate system, but parallel to the direction of the thickness for the generality. Apparently, when the imaging point of the object is obtained, the ray to this object can be computed.

Suppose that X_0 is the imaging point on the image plane, which can be easily obtained by mature methods of image processing. The ray $\mathbf{r}_0$ can then be denoted by the following equation:

$$\mathbf{r}_0 = \frac{\overrightarrow{OX_0}}{\left|\overrightarrow{OX_0}\right|}. \tag{3.2}$$

Further, the point on the first refractive surface can be expressed as

$$X_1 = \frac{d_0}{\mathbf{r}_0 \cdot \mathbf{n}_\pi} \mathbf{r}_0. \tag{3.3}$$

Agrawal et al. have proved that all segments of the ray from the optical center to the object, together with the unit normal vector $\mathbf{n}_\pi$, lie in a common plane, the plane of refraction (POR) [24], which can be presented by

$$(\gamma \mathbf{r}_i + \eta \mathbf{r}_{i+1}) \cdot (\mathbf{r}_i \times \mathbf{n}_\pi) = 0, \tag{3.4}$$

where $i \in [0, N-1]$, γ and η are two non-zero real numbers.

Under the constraint of POR, the ray through the refractive surface satisfies the linear combination of the normal vector with the direction vector of the previous ray segment as follows:

$$\mathbf{r}_{i+1} = \alpha_i \mathbf{r}_i + \beta_i \mathbf{n}_\pi. \tag{3.5}$$

In Figure 3.2b, notice that owing to the different comparison between μ_i and μ_{i+1}, the values of α_i and β_i are different.

If $\mu_i \le \mu_{i+1}$,

$$
\begin{cases}
\alpha_i = \dfrac{\mu_i}{\mu_{i+1}} \\[3mm]
\beta_i = \dfrac{\mu_i}{\mu_{i+1}} \mathbf{r}_i \cdot \mathbf{n}_\pi - \sqrt{1 - \left(\dfrac{\mu_i}{\mu_{i+1}}\right)^2 [1 - (\mathbf{r}_i \cdot \mathbf{n}_\pi)^2]}
\end{cases}
. \tag{3.6}
$$

If $\mu_i > \mu_{i+1}$,

$$
\begin{cases}
\alpha_i = \dfrac{\mu_i}{\mu_{i+1}} \\[3mm]
\beta_i = \sqrt{1 - (\dfrac{\mu_i}{\mu_{i+1}})^2 [1 - (\mathbf{r}_i \cdot \mathbf{n}_\pi)^2]} - \dfrac{\mu_i}{\mu_{i+1}} \mathbf{r}_i \cdot \mathbf{n}_\pi
\end{cases}
. \tag{3.7}
$$

Now, we have each point on the refractive surface

$$
X_k = \sum_{i=0}^{k-1} \frac{d_i}{\mathbf{r}_i \cdot \mathbf{n}_\pi} \mathbf{r}_i \qquad k \in [1, N-1]. \tag{3.8}
$$

Based on equations (3.2)–(3.8), the point on the final refractive surface X_N and the direction of the final segment $\mathbf{r}_N$ are obtained by computing layer by layer. Apparently, the object is on the ray directed by $\mathbf{r}_N$, but its position cannot be determined with only one camera. Hence, the akin triangulation is necessary.

3.4 AKIN TRIANGULATION AND REFRACTIVE CONSTRAINT

3.4.1 Akin Triangulation

On account of the refractive camera model in Section 3.3, the traditional triangulation based on the linear relationship of the multiple view geometry cannot be appropriate for getting the accurate 3D position information of underwater objects [20, 21]. Sequentially, without the fundamental matrix constraint, many classical binocular methods are out of operation. In this section, the akin triangulation is applied to improve underwater binocular measurement. Because the combination of the ray from different cameras is not triangular, this process is regarded as the akin triangulation.

As shown in Figure 3.3a, M cameras are located at the optical center O_i. Note that $\{X_{0,1}, X_{0,2}, \ldots, X_{0,M}\}$ is a set of corresponding imaging points of the same object. $\{X_{WG,1}, X_{WG,2}, \ldots, X_{WG,M}\}$ is a set of intersecting points on the water-to-glass surface. $\{\mathbf{r}_{P,1}, \mathbf{r}_{P,2}, \ldots, \mathbf{r}_{P,M}\}$ is a set of the direction vector of the final ray segment to the object. Based on Section 3.3, we can obtain the value of $X_{WG,i}$ and $\mathbf{r}_{P,i}$ easily in different camera coordinate references.

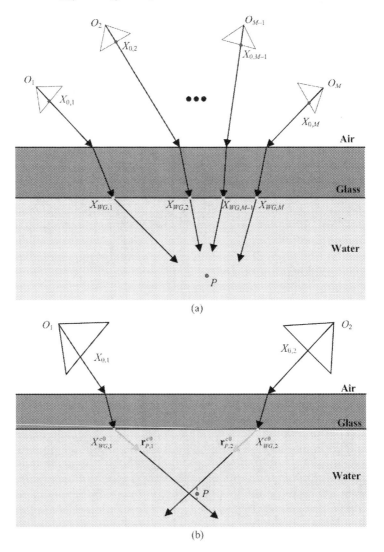

FIGURE 3.3   The description of the akin triangulation. (a) Multiple camera system case. (b) UBVMS case.

Here, the camera O_1 is defined as the basic camera, $\mathbf{R}_i$, represents the ith camera rotation matrix, and $\mathbf{T}_i$ denotes the ith camera translation vector. Then, we translate intersecting points and direction vector to the basic camera coordinate system by the following equations:

$$X_{WG,i}^{c0} = \mathbf{R}_i X_{WG,i} + \mathbf{T}_i, \tag{3.9}$$

$$\mathbf{r}_{P,i}^{c0} = \mathbf{R}_i \mathbf{r}_{P,i} + \mathbf{T}_i. \tag{3.10}$$

In this multiple camera system, the position of the object P can be determined by the crossover point of all rays, which is similar to the traditional triangulation. Considering the system errors, these rays cannot intersect each other exactly. We can seek the object points by minimizing the sum of the distance from P to all rays. Next, $L(A,\mathbf{b})$ is defined as a straight line through point A with a direction vector $\mathbf{b}$, and $Dis(P,L)$ denotes the distance from point P to the straight line $\mathbf{L}$. Therefore, the process of seeking the actual position of the object is described as

$$P = \arg\min_{p} \sum_{i=1}^{M} Dis(p, \mathbf{L}(X_{WG,i}^{c0}, \mathbf{r}_{P,i}^{c0})). \tag{3.11}$$

In the UBVMS, just two cameras are installed, as shown in Figure 3.3b. Consequently, the equation (3.11) can be simplified by using the midpoint of the perpendicular to approach the object point. As is given,

$$X_{WG,1}^{c0} = \begin{pmatrix} x_{WG,1} \\ y_{WG,1} \\ z_{WG,1} \end{pmatrix} \qquad X_{WG,2}^{c0} = \begin{pmatrix} x_{WG,2} \\ y_{WG,2} \\ z_{WG,2} \end{pmatrix}, \tag{3.12}$$

$$\mathbf{r}_{P,1}^{c0} = \begin{pmatrix} m_{P,1} \\ n_{P,1} \\ p_{P,1} \end{pmatrix} \qquad \mathbf{r}_{P,2}^{c0} = \begin{pmatrix} m_{P,2} \\ n_{P,2} \\ p_{P,2} \end{pmatrix}. \tag{3.13}$$

Then, the straight line $\mathbf{L}(X^{c0}_{WG,1}, \mathbf{r}^{c0}_{P,1})$ and $\mathbf{L}(X^{c0}_{WG,2}, \mathbf{r}^{c0}_{P,2})$ are denoted separately in the following equations:

$$\frac{x - x_{WG,1}}{m_{P,1}} = \frac{y - y_{WG,1}}{n_{P,1}} = \frac{z - z_{WG,1}}{p_{P,1}}, \tag{3.14}$$

$$\frac{x - x_{WG,2}}{m_{P,2}} = \frac{y - y_{WG,2}}{n_{P,2}} = \frac{z - z_{WG,2}}{p_{P,2}}. \tag{3.15}$$

Let t_1 and t_2 are two intermediate variables, it follows:

$$t_1 = \frac{\begin{vmatrix} n_{P,1}p_{P,2} - n_{P,2}p_{P,1} & x_{WG,2} - x_{WG,1} & m_{P,2} \\ p_{P,1}m_{P,2} - p_{P,2}m_{P,1} & y_{WG,2} - y_{WG,1} & n_{P,2} \\ m_{P,1}n_{P,2} - m_{P,2}n_{P,1} & z_{WG,2} - z_{WG,1} & p_{P,2} \end{vmatrix}}{\begin{vmatrix} n_{P,1}p_{P,2} - n_{P,2}p_{P,1} & m_{P,1} & m_{P,2} \\ p_{P,1}m_{P,2} - p_{P,2}m_{P,1} & n_{P,1} & n_{P,2} \\ m_{P,1}n_{P,2} - m_{P,2}n_{P,1} & p_{P,1} & p_{P,2} \end{vmatrix}},$$

$$t_2 = -\frac{\begin{vmatrix} n_{P,1}p_{P,2} - n_{P,2}p_{P,1} & m_{P,1} & x_{WG,2} - x_{WG,1} \\ p_{P,1}m_{P,2} - p_{P,2}m_{P,1} & n_{P,1} & y_{WG,2} - y_{WG,1} \\ m_{P,1}n_{P,2} - m_{P,2}n_{P,1} & p_{P,1} & z_{WG,2} - z_{WG,1} \end{vmatrix}}{\begin{vmatrix} n_{P,1}p_{P,2} - n_{P,2}p_{P,1} & m_{P,1} & m_{P,2} \\ p_{P,1}m_{P,2} - p_{P,2}m_{P,1} & n_{P,1} & n_{P,2} \\ m_{P,1}n_{P,2} - m_{P,2}n_{P,1} & p_{P,1} & p_{P,2} \end{vmatrix}}.$$

Then, the coordinate of the object point is derived by

$$P = \begin{pmatrix} \dfrac{x_{WG,1} + m_{P,1}t_1 + x_{WG,2} + m_{P,2}t_2}{2} \\[2mm] \dfrac{y_{WG,1} + n_{P,1}t_1 + y_{WG,2} + n_{P,2}t_2}{2} \\[2mm] \dfrac{z_{WG,1} + p_{P,1}t_1 + z_{WG,2} + p_{P,2}t_2}{2} \end{pmatrix}. \tag{3.16}$$

Finally, we obtain the complete nonlinear geometrical relationship by the akin triangulation, which forms the mathematical basis of the calibration algorithm.

3.4.2 Refractive Surface Constraint

It is inspired by the akin triangulation that all of the intersecting points on the glass-to-air or the water-to-glass surface are in the common plane as the following equation:

$$\mathbf{n}_\pi \cdot \sum_{i=1}^{M-1} \left(X_{WG,i}^{c0} - X_{WG,i+1}^{c0} \right) = 0. \tag{3.17}$$

For a binocular camera system, the constraint is simplified by,

$$\mathbf{n}_\pi \cdot \left(X_{WG,1}^{c0} - X_{WG,2}^{c0} \right) = 0. \tag{3.18}$$

The refractive surface constraint is quite necessary because calibrated parameters from the undermentioned optimization algorithm will probably compute two parallel glass-to-air or water-to-glass surfaces due to the multiple solutions of the nonlinear refractive relationship equation. Therefore, with the refractive surface constraint, there are only a single-layer glass-to-air surface and a single-layer water-to-glass surface in the result of optimization.

3.5 CALIBRATION ALGORITHM

According to previous analysis of the refractive camera model and akin triangulation for UBVMS, there is a nonlinear relationship among the object coordinate, the normal of refractive surfaces $\mathbf{n}_\pi$, the vertical distance from optical center to the first refractive surface d_L and d_R, and the thickness of the glass h. Note that $\mathbf{n}_\pi$, d_L, d_R, and h are defined as the binocular housing parameters. So the object point can be expressed by a nonlinear function:

$$Obj^i = f(X_L^i, X_R^i, \mathbf{n}_\pi, d_L, d_R, h). \tag{3.19}$$

It is not easy to solve the nonlinear refractive relationship equation directly. Instead, adopting optimization methods to explore binocular housing parameters of UBVMS is a better way. Therefore, we propose a novel usage of the checkerboard to set the optimization goal and employ NSGA-II to achieve it.

3.5.1 A Novel Usage of Checkerboard

The checkerboard is widely used for camera calibration recently [30–33], related to the homography between the checkerboard and its image [3]. The homograpy originates from equation (3.1), based on the pinhole model, and is inapplicable to UBVMS. Next, a novel usage of checkerboard for underwater calibration is introduced.

As can be observed from Figure 3.4, the distance between two corners and the relative location among three corners can be easily determined in advance. $C_{p,q}$ denotes the corner in checkerboard pattern coordinate system, and $C_{p,q}^c$ denotes the corner in the camera coordinate system computed according to the proposed refractive camera model. Note that the subscript of C signifies the coordinate in checkerboard pattern coordinate system. If the calibration is accurate, the following equations can be established:

$$\left|\overrightarrow{C_{p,q}^c C_{m,n}^c}\right| = \left|\overrightarrow{C_{p,q} C_{m,n}}\right| = w\sqrt{(m-p)^2 + (n-q)^2}, \tag{3.20}$$

$$\overrightarrow{C_{p,q}^c C_{m,n}^c} \cdot \overrightarrow{C_{p,q}^c C_{s,t}^c} = \overrightarrow{C_{p,q} C_{m,n}} \cdot \overrightarrow{C_{p,q} C_{s,t}}. \tag{3.21}$$

where $\overrightarrow{*_1 *_2}$ and $\left|\overrightarrow{*_1 *_2}\right|$ denote a vector from corner point $*_1$ to corner point $*_2$ and its distance, respectively.

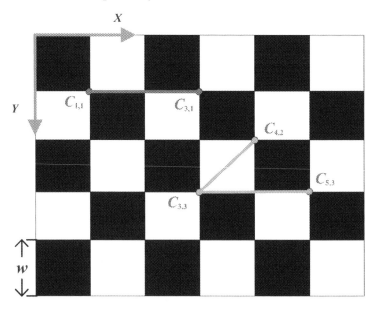

FIGURE 3.4 The relative position between corners.

Assuming the size of the checkerboard is $U \times V$, we propose three optimization goals for the calibration process.

The first goal is to minimize the distance difference:

$$(\mathbf{n}_\pi, d_L, d_R, h) = \arg\min_{\mathbf{n}_\pi, d_L, d_R, h} F_1,$$

(3.22)

$$F_1 = \sum_{i=1}^{U-1} \sum_{j=1}^{V-1} \left| \left| \overrightarrow{C_{i,j}^c C_{i+1,j}^c} \right| + \left| \overrightarrow{C_{i,j}^c C_{i,j+1}^c} \right| - 2w \right|.$$

This equation means that for each corner point, the measured distance from it to its right or down adjacent point is close to the true value.

The second goal is to minimize the vertical direction difference:

$$(\mathbf{n}_\pi, d_L, d_R, h) = \arg\min_{\mathbf{n}_\pi, d_L, d_R, h} F_2,$$

(3.23)

$$F_2 = \sum_{i=1}^{U-1} \sum_{j=1}^{V-1} \left| \frac{\overrightarrow{C_{i,j}^c C_{i+1,j}^c} \cdot \overrightarrow{C_{i,j}^c C_{i,j+1}^c}}{\left| \overrightarrow{C_{i,j}^c C_{i+1,j}^c} \right| \left| \overrightarrow{C_{i,j}^c C_{i,j+1}^c} \right|} \right|.$$

This optimal goal expresses that the four angles at each corner point are right angles.

The third goal is to minimize the parallel direction difference:

$$(\mathbf{n}_\pi, d_L, d_R, h) = \arg\min_{\mathbf{n}_\pi, d_L, d_R, h} F_3,$$

$$F_3 = \sum_{i=1}^{U-1} \sum_{j=1}^{V-1} \left| F_3^1 + F_3^2 - 2 \right|,$$

(3.24)

$$F_3^1 = \frac{\overrightarrow{C_{i,j}^c C_{i+1,j}^c} \cdot \overrightarrow{C_{1,1}^c C_{V,1}^c}}{\left| \overrightarrow{C_{i,j}^c C_{i+1,j}^c} \right| \left| \overrightarrow{C_{1,1}^c C_{V,1}^c} \right|}, F_3^2 = \frac{\overrightarrow{C_{i,j}^c C_{i,j+1}^c} \cdot \overrightarrow{C_{1,1}^c C_{1,U}^c}}{\left| \overrightarrow{C_{i,j}^c C_{i,j+1}^c} \right| \left| \overrightarrow{C_{1,1}^c C_{1,U}^c} \right|}.$$

Note that $F_3^1 = 1$ means that the vector from a corner point to its right adjacent corner point is parallel to X axis direction of the checkerboard. Similarly, $F_3^2 = 1$ denotes that the vector from a corner point to its down adjacent corner point is parallel to Y axis direction of the checkerboard.

In this fashion, combining the three optimization goals, this calibration process is regarded as a multi-objective optimization problem as

$$(\mathbf{n}_\pi, d_L, d_R, h) = \underset{\mathbf{n}_\pi, d_L, d_R, h}{\arg\min}\{F_1, F_2, F_3\}. \tag{3.25}$$

3.5.2 Analysis of the Binocular Housing Parameters

In the aforementioned analysis, the number of binocular housing parameters is six. Note that the two mathematical and geometrical relationships can simplify the process of optimization. The first one is the constraint of the unit normal vector as follows:

$$\mathbf{n}_\pi(z) = \sqrt{1 - \mathbf{n}_\pi(x)^2 - \mathbf{n}_\pi(y)^2}. \tag{3.26}$$

The second one is the geometrical relationship of d_L and d_R:

$$d_R = d_L - \mathbf{T} \cdot \frac{\mathbf{T} \cdot \mathbf{n}_\pi}{|\mathbf{T} \cdot \mathbf{n}_\pi|}, \tag{3.27}$$

where $\mathbf{T}$ denotes the translation vector between the left camera and the right camera. In this situation, the number of housing parameters declines to four, only including $\mathbf{n}_\pi(x)$, $\mathbf{n}_\pi(y)$, d_L, and h.

In addition, the search space of binocular housing parameters can be confirmed in advance. First of all, the refractive surface constraint mentioned in Section 3.4 is applied to restrict the search space. In practice, the image plane of the binocular camera and the surface of the waterproof house are installed as parallel as possible for a stable working circumstance, which restricts the value of $\mathbf{n}_\pi(x)$ and $\mathbf{n}_\pi(y)$ to vary in the range of $[-0.3, 0.3]$. The variation range of d_L and h can be estimated approximately by manual measurement.

3.5.3 NSGA-II Algorithm

The genetic algorithm proposed in [34] is a parallel global search method for optimization, which can accumulate the information related to the search space automatically and get the optimum relation by adaptively controlling the searching process [35]. NSGA-II [36], proposed by Deb et al., is a non-dominated sorting genetic algorithm for multi-objective optimization. NSGA-II proposes a new strategy and a new arithmetic operator by improving the original NSGA [37] algorithm: The fast non-dominated sorting approach and the crowded comparison operator. In literature, many works [38,39] about optimization based on genetic have been done. And all these achievements demonstrate that the genetic algorithm and its modified methods are feasible for optimal design.

The principle of NSGA-II shown in Figure 3.5 can be described as follows. The offspring population Q_t, which is obtained after the fast non-dominated sorting approach employed on the parent population P_i, combines with P_t to constitute a new population $R_i(R_t = P_t \cup Q_t)$ with the size of 2N. When the non-domination sorting is conducted on the population R_t, the crowded comparison operator will be used to compare the individuals with the equative front calculated by fast sorting. Therefore, a new population P_{t+1} will be generated by eliminating the improper solution, which is the solution overflow of N. By looping and eliminating continuously, a group of optimal solutions will be obtained which are in the Pareto-optimal frontiers. In practice, one or several solutions should be chosen from Pareto-optimal solutions as the optimal solution for solving the multi-objective optimization problems. Figure 3.6 shows the Pareto-optimal solutions of the proposed three objective goals.

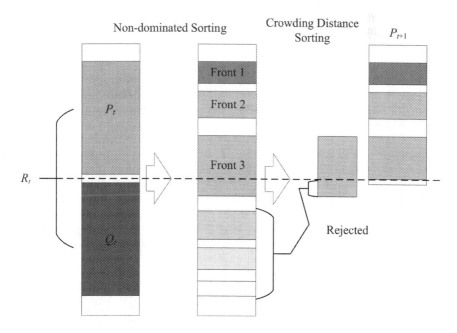

FIGURE 3.5   Procedure of NSGA-II.

Algorithm 1 NSGA-II based underwater calibration algorithm

Input: checkerboard images in air *CA*, checkerboard images underwater *CU*
Output: calibrated housing parameters *Hout*
1: [*Intrinsic, Extrinsic, Distortion*] = CalibrateAir(*CA*);
2: Cornersimg = DetectCorners(*Intrinsic, Extrinsic, Distortion, CU*);
3: set the total generation *nGen*;
4: set the population quantity *nPop*;
5: initial housing parameters *H*;
6: *i* = 0;
7: **for** *i* < *nPop* **do**
8:          *H(i)* = random();
9: *Gen* = 1;
10: set optimal goals *F1, F2, F3*;
11: **for** *Gen* < *nGen* **do**
12:               *H.Cornerspos* = RefractiveModel (*Cornersimg, H*);
13:               **while** RefractiveSurfaceConstraint (*H.Cornerspos*) == 0 **do**
14:     *H* = CrossingOver(*H*);
15:   *H* = Mutation(*H*);
16:     *H.Cornerspos* = RefractiveModel (*Cornersimg, H*);
17:               *H* = NSGA-II(*H, F1, F2, F3*);
18: **return** one solution of *H*;

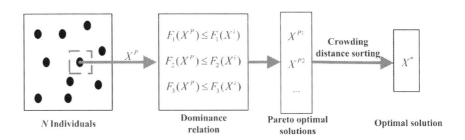

N Individuals   Dominance relation   Pareto optimal solutions   Optimal solution

FIGURE 3.6   Pareto-optimal solutions of the proposed three objective functions F_1, F_2, and F_3.

3.5.4 Process of the Calibration Algorithm

The process of our calibration algorithm for the UBVMS is shown in Algorithm 1. First of all, the binocular camera is calibrated by a methodology in the air to obtain its intrinsic matrix, extrinsic matrix, and distortion parameters, which are invariant for underwater circumstance. Secondly, we obtain images of underwater checkerboard and remove the distortion, followed by the detection of checkerboard corners. Then NSGA-II parameters *nGen* and *nPop* are set, each random group of binocular housing parameters is initialed as the individual in the population. With respect to each individual, the 3D positions of corners are computed according to the refractive models and akin triangulation as equation (3.19), which must conform to the refractive surface constraint as equation (3.18). If one of the individuals is out of the constraint, the crossing over and mutation operation are executed. Then, we solve the multi-objective optimization as equation (3.25) by NSGA-II to update an appropriate group of housing parameters. Finally, when the present generation *Gen* reaches the total generations *nGen*, the optimization is finished. One of the solutions in *H* is picked out as the calibrated binocular housing parameter.

3.6 EXPERIMENTS AND RESULTS

3.6.1 Experimental Setup

To accomplish the following experiments, we choose a tank with the dimensions of 120 cm long, 80 cm wide, and 80 cm deep. As shown in Figure 3.7, a binocular camera connected to the computer and protected

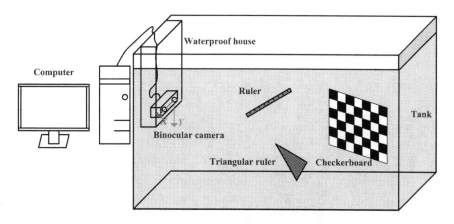

FIGURE 3.7   Schematic diagram of an experimental setup.

by a glass waterproof is installed at the left side of the tank. A checker-board, a ruler, and a triangular ruler are employed to demonstrate different performances. The whole measurement experiments refer to the left camera coordinate system. Finally, the experimental results are numerically computed in Matlab.

3.6.2 Results of Calibration

First of all, we obtain the intrinsic matrix, extrinsic matrix, and distortion parameters in the air. The results of calibration in the air are shown as:

$$\mathbf{I}_{\text{left}} = \begin{pmatrix} 696.359 & 0 & 659.513 \\ 0 & 695.987 & 334.085 \\ 0 & 0 & 1 \end{pmatrix}$$

$$\mathbf{I}_{\text{right}} = \begin{pmatrix} 694.535 & 0 & 657.004 \\ 0 & 694.155 & 347.721 \\ 0 & 0 & 1 \end{pmatrix}$$

$$\mathbf{Ex} = \begin{pmatrix} 1.000 & -0.002 & -0.004 & 119.741 \\ 0.002 & 0.999 & -0.014 & -0.327 \\ 0.0037 & 0.014 & 0.999 & 1.154 \\ 0 & 0 & 0 & 1 \end{pmatrix}$$

$$\mathbf{k}_{\text{left}} = [-0.150 \ -0.011] \quad \mathbf{k}_{\text{right}} = [-0.157 \ 0.010]$$

where $\mathbf{I}_{\text{left}}$ and $\mathbf{I}_{\text{right}}$ denote the intrinsic matrix of the left camera and the intrinsic matrix of the right camera, respectively, $\mathbf{Ex}$ is the extrinsic matrix, and $\mathbf{k}_{\text{left}}$ and $\mathbf{k}_{\text{right}}$ denote the radial distortion vector of the binocular camera.

Two checkerboards in different sizes are chosen in this step. One is 8×16 grids with 25 mm side length, and the other is 6×9 grids with 30 mm side length. Images of the former are used to calibrate the parameters, and images of the latter are utilized in a test set to verify the calibration result. The number of images for each checkerboard is 30. Note that the variation of the calibration parameters can be estimated by a prior rough measurement. In our experimental setup, $\mathbf{n}_\pi(x)$ and $\mathbf{n}_\pi(y)$ vary in the range of $[-0.3, 0.3]$, h varies in the range of $[-6.0, 7.0]$, and d_L varies in the range of $[-14.0, 15.0]$. In the NSGA-II procedure, $nGen$ is set to 100 and $nPop$ is set to 50. Note that the crossover rate is empirically set to 0.8 and the mutation rate is set to 0.1. The appropriate crossover and mutation rate will increase the population diversity in order to improve the convergence speed and avoid local optimum solutions. In addition, the Partial-Mapped Crossover (PMX) and Simple Mutation are separately adopted for crossover and mutation. After calibration, we get the result as $\mathbf{n}_\pi(x) = 0.0010$, $\mathbf{n}_\pi(y) = 0.0998$, $d_L = 14.9878$, and $h = 6.5325$.

Shortis et al. indicated that the accuracy of the calibration parameters would be enhanced if the attitudes of checkerboard were various and the calibration range was extensive enough, for a high level of redundant information, provided by many target image observations on many exposures, could eliminate the outliers [10]. Therefore, for calibrating the UBVMS precisely and reliably by the proposed algorithm, the attitudes and the calibration range of the checkerboard should meet this rule.

In equation (3.25), there are three optimization goals including the distance difference, the vertical direction difference, and the parallel direction difference, which can be regarded as the evaluation index on the test set. The distance difference is the error between the measured value and the real value of the length of each grid side. The vertical direction difference and parallel direction difference are computed referring to equations (3.23) and (3.24). Therefore, the average differences in distance, vertical direction, and parallel direction are computed in each image of the test set shown in Figure 3.8. Noticeably, each average difference distributes in a small range around zero. Furthermore, the measurement accuracy rate of distances on checkerboard reaches 98.5% by the radial lens distortion correction [10], whereas the measurement accuracy rate by the proposed method is up to 98.9%. Therefore, the calibration results are satisfactory.

It is necessary to verify the feasibility of the proposed calibration algorithm with different calibration attempts such as different underwater environments, different camera performances, and different accuracies of the checkerboard corner detection algorithms. The primary factor for impact on the calibration results is the difference in pixel coordinates of checkerboard corners with different calibration attempts. However, limited to the existing experimental condition, the real data from different calibration attempts are difficult to obtain. In this situation, referring to

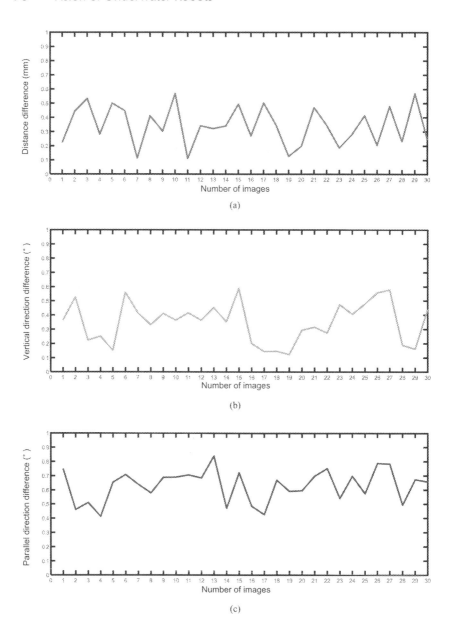

FIGURE 3.8   Evaluation index on the test set. (a) The average distance difference of each checkerboard image on the test set. (b) The average vertical difference of each checkerboard image on the test set. (c) The average parallel difference of each checkerboard image on the test set.

the simulative validation method of Telem et al. [26], an additive normally distributed random shift noise with $\sim 2N(0,\sigma_0^2)$ is applied to each corner pixel coordinate to simulate the shift of pixel coordinates. Along with σ_0 from 0 to 1.2 by step of 0.3, the calibration parameters results are shown in Table 3.1. The stable results of calibration parameters demonstrate that the proposed calibration algorithm is robust for different attempts.

3.6.3 Experiments on Position Measurement

The experiment is designed as follows to demonstrate the accurate position measurement. Firstly, the real location of the checkerboard is obtained by the binocular triangular algorithm in air, when the tank is without water. Then, the tank is filled with water, and the location of the checkerboard is fixed and unchanged. In the underwater circumstance, we choose three methods to calibrate the binocular camera and then measure the position of the checkerboard. **Method I** is the direct binocular calibration algorithm with intrinsic matrix, extrinsic matrix, and the distortion parameters calibrated in air. **Method II** is the binocular calibration algorithm with intrinsic matrix, extrinsic matrix, and the distortion parameters calibrated underwater without regard to the explicit refractive model [16, 17]. **Method III** is proposed in this chapter considering the refractive camera model. The ground truth is obtained by the couple of paired images in the air in a well-known binocular measurement methodology.

The position coordinates measurement results, in the left camera reference system, are shown as given in Figure 3.9. **Method I** is terrible, for the surface of the checkerboard is bent, and the measured position is significantly different from the **Ground Truth**. **Method II** is closer to the **Ground Truth** than **Method I** but an unacceptable error still exists. Compared to **Method I** and **Method II**, it is apparent that **Method III** has a better position measurement performance with an accurate shape of the checkerboard and accurate positions of corners.

TABLE 3.1 Calibration results with noises

σ_0	$n_\pi(x)$	$n_\pi(y)$	d_L	h
0	0.0010	0.0998	14.9878	6.5325
0.3	0.0013	0.0994	14.9862	6.5313
0.6	0.0014	0.1011	14.9754	6.5210
0.9	0.0003	0.0960	14.9823	6.5721
1.2	0.0025	0.1140	14.8865	6.3549

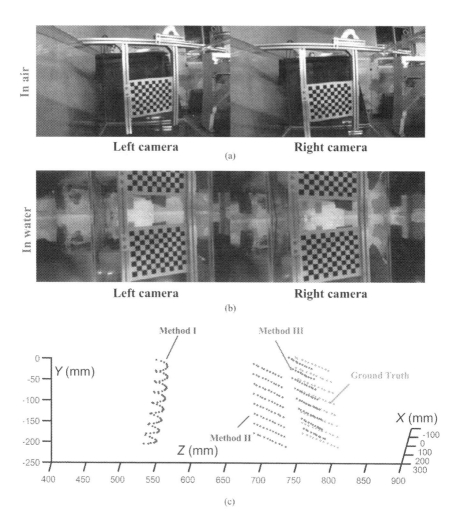

FIGURE 3.9 Results of position measurement. (a) Image of the fixed checkerboard in air. (b) Image of the fixed checkerboard in water. (c) The results of position measurement with the left camera coordinate system as the reference.

For quantitative analysis, an error index *Err* is used to evaluate the results of the three methods:

$$Err_{\text{Method}^*} = \frac{\sum_{i=1}^{nCorners} (Pos_{i,\text{Method}^*} - Pos_{i,\text{GT}})}{nCorners}, \tag{3.28}$$

where *nCorners* is the number of corners of the checkerboard, Pos_{i,Method^*} denotes the measured position of a corner according to **Method** *, and

$Pos_{i,\text{GT}}$ is the ground truth position. The computed results are listed in Table 3.2, which intuitively indicate that the binocular camera calibrated by **Method III** can measure the position of objects underwater more accurately and reliably.

3.6.4 Experiments on Position Measurement

The binocular camera, separately calibrated by **Method II** and **Method III**, measures the length of a ruler and the angle of a triangular ruler underwater. The ruler is marked by two points, between which the distance is 300 mm, and the triangular ruler has a 30° angle and a 60° angle. We capture images of the ruler and the triangular ruler at different attitudes underwater, as shown in Figure 3.10. Corresponding length and angle measurement results are listed in Table 3.3.

For underwater measurements, **Method II** performs worse. Inversely, calibrated by **Method III**, the binocular camera measures the length at the average accuracy up to 97.7% and measures the angle at about 97.5%. In the case of underwater short distance operations requiring higher precision, the accuracy is sufficient.

3.6.5 Discussion

According to the experimental results, the UBVMS calibrated by our algorithm can measure the position, length, and angle accurately. The binocular camera is calibrated by **Method III** to obtain the binocular housing parameters, overcoming the problem of image change caused by refractions.

Compared to **Method I** and **Method II**, **Method III** is much more logical and considerate. In **Method I**, the refractive camera model is considered into calibration in order to improve the measurement accuracy. **Method II** takes refractions into consideration, but the refractive relationship is not explicit and is regarded as linear, leading to the unreliable measurement. It is apparent that the proper refractive camera model is

TABLE 3.2   Error of position measurement

Index	Value (mm)
*Err*Method I	224.25
*Err*Method II	69.35
*Err*Method III	17.89

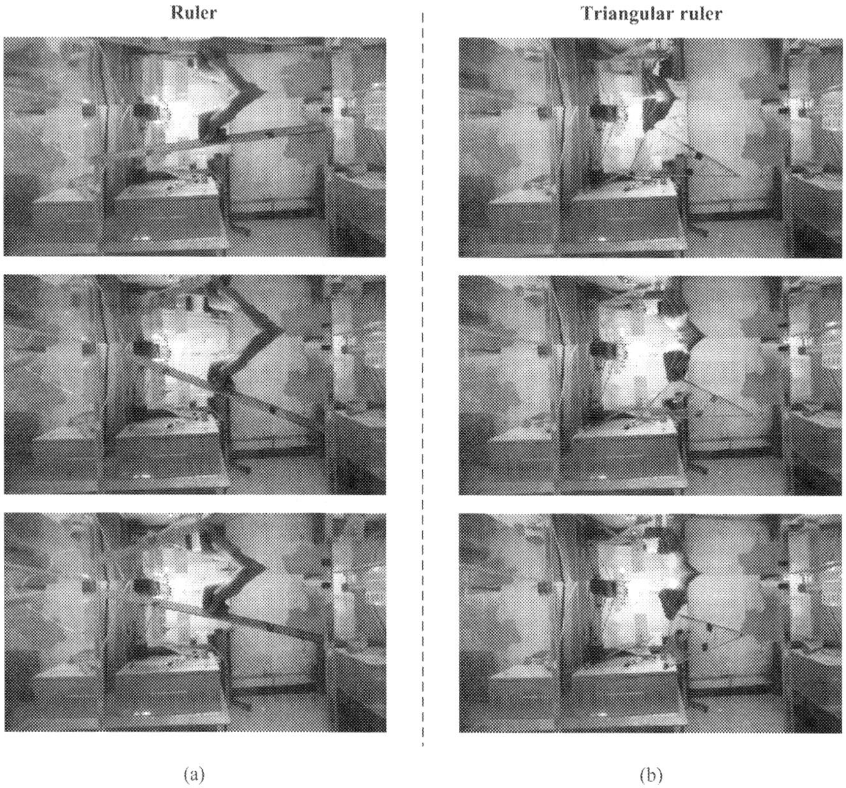

Ruler | Triangular ruler

(a)   (b)

FIGURE 3.10   Experiments on length and angle measurement. Note that the ruler (a) and the triangular rulers (b) are put at different attitudes to prove the robustness of the method.

TABLE 3.3   Results of length and angle measurement

Calibration method		Method II	Method III
Length (300 mm)	Average value	282.88 mm	306.81 mm
	Accuracy	94.3%	97.7%
Degree (30°)	Average value	26.06°	29.42°
	Accuracy	86.7%	98.1%
Degree (60°)	Average value	72.04°	58.48°
	Accuracy	79.9%	97.5%

essential for accurate measurement. Note that the key reason for the accurate measurement is to compute accurate position coordinates of objects. The three optimal goals derived from the novel usage of the checkerboard are organized to represent precise positions of checkerboard corners, which urges the measurement accuracy.

Solving the nonlinear refractive relationship between the objects and their corresponding points on the image plane directly is complicated depending on complex mathematical analysis. Compared to the calibration methodologies presented in [23–25,27,29], **Method III** uses a multiobjective optimization to avoid the complex mathematical analysis of refractive relationship. The intelligent optimization algorithm NSGA-II is a sagacious choice.

Furthermore, the motivation of the proposed algorithm is to provide support technology for underwater robot operations in the future. Similarly, the time-of-flight, one of the most recent calibration methods, can be applied to underwater robot operations [28]. Under the background of underwater robot application, the proposed algorithm is compared with the time-of-flight in measuring accuracy, measuring distance, geometrical conditions, auxiliary means, and robotic applications. On account of the difficulty in reproducing the same experimental setup, a direct quantitative comparison with the time-of-flight is impossible. However, major differences are presented in Table 3.4. Note that the proposed algorithm has a slightly lower accuracy than the time-of-flight, but it is enough for underwater robot operations. Redundant design of the mechanical structure will compensate for the measurement errors. In addition, it is practically difficult to keep the optical axis direction vertical to refractive surface. Therefore, the random optical axis direction is more general for different applications. Calibration only by a typical checkerboard can be implemented easily in different field environments. In addition, the proposed algorithm is suitable for various binocular cameras, but the time-of-flight demands cameras equipped with infrared ray like Kinect. If some underwater robots have limited space to carry the cameras equipped with infrared ray, the proposed method will be more appropriate for UBVMS. In conclusion, the proposed algorithm has wider application scenarios than time-of-flight.

TABLE 3.4   Comparison of proposed algorithm with time-of-flight

Aspect	Proposed algorithm	Time-of-flight
Measuring accuracy	>97.5%	>99.0%
Measuring distance	<1,000 mm	350 mm–650 mm
Geometrical conditions	Random optical axis direction	Optical axis is vertical with refractive surface
Auxiliary means	A typical checkerboard	Infrared ray
Robotic applications	Robotic grasp and follow	Underwater scene reconstruction

Although the feasibility of the calibration proposed in this chapter is demonstrated by experiments, there are still two potential problems. Firstly, the refractive camera model is based on the ideal underwater model, ignoring the light scattering, light absorption, and another phenomenon of light distortion. So the measurement errors are ineradicable by this calibration algorithm. To solve this problem, the influence of light distortion will be under consideration for modeling the refraction. In this situation, a prime advantage of utilizing the optimization algorithm for the calibration will appear, because the refractive relationship cannot be described by simple mathematical equations. Secondly, the NSGA-II algorithm has the premature convergence problem, if the search space of binocular housing parameters is unreasonable. This problem can be overcome by modifying the multi-objective optimization.

3.7 CONCLUSION AND FUTURE WORK

This chapter has proposed a feasible calibration algorithm for UBVMS. Firstly, the general refractive camera model is established and binocular housing parameters are introduced. Next, the akin triangulation is described to determine the position of objects. Derived from the akin triangulation, the refractive surface constraint is to ensure that multiple solutions are avoided. To obtain the housing parameters, a new usage of the checkerboard is proposed to set three optimization goals instead of the direct solving of nonlinear refractive relationship. By means of NSGA-II, the binocular housing parameters are calibrated. Finally, experimental results demonstrate the feasibility of the calibration algorithm.

In the future, the calibration algorithm will be improved for the binocular camera, considering the light distortion and the optimization premature convergence problem, to measure position coordinates more precisely. Then, the underwater 3D reconstruction will be further investigated. When the relevant underwater vision technologies are developed, the UBVMS can be embedded in the ROV or robotic fish for underwater operations.

REFERENCES

1. M. Palmese and A. Trucco, "Acoustic imaging of underwater embedded objects: Signal simulation for 3-D sonar instrumentation," *IEEE Trans. Instrum. Meas.*, vol. 55, no. 4, pp. 1339–1347, 2006.

2. Trucco, M. Palmese, and S. Repetto, "Devising an affordable sonar system for underwater 3-D vision," *IEEE Trans. Instrum. Meas.*, vol. 57, no. 10, pp. 2348–2354, 2008.
3. Z. Zhang, "A flexible new technique for camera calibration," *IEEE Trans. Pattern Anal. Mach. Intell.*, vol. 22, no. 11, pp. 1330–1334, 2000.
4. R. Y. Tsai, "A versatile camera calibration technique for high-accuracy 3D machine vision metrology using off-the-shelf TV cameras and lenses," *IEEE J. Rob. Autom.*, vol. 3, no. 4, pp. 323–344, 1987.
5. H. Bacakoglu and M. S. Kamel, "A three-step camera calibration method," *IEEE Trans. Instrum. Meas.*, vol. 46, no. 5, pp. 1165–1172, 1997.
6. J. L. L. Galilea, J. Lavest, C. A. L. Vazquez, A. G. Vicente, and I. B. Munoz, "Calibration of a high-accuracy 3-D coordinate measurement sensor based on laser beam and CMOS camera," *IEEE Trans. Instrum. Meas.*, vol. 58, no. 9, pp. 3341–3346, 2009.
7. Sedlazeck and R. Koch, "Perspective and non-perspective camera models in underwater imaging-overview and error analysis," in *Outdoor and Large-Scale Real-World Scene Analysis*. Berlin, Germany: Springer, 2012, pp. 212–242.
8. N. Boutros, M. R. Shortis, and E. S. Harvey, "A comparison of calibration methods and system configurations of underwater stereo-video systems for applications in marine ecology," *Limnol. Oceanogr. Methods*, vol. 13, no. 5, pp. 224–236, 2015.
9. M. Shortis, and E. H. D. Abdo, "A review of underwater stereo-image measurement for marine biology and ecology applications," in *Oceanography and Marine Biology*. Boca Raton, FL, USA: CRC Press, 2016, pp. 269–304.
10. M. Shortis, "Calibration techniques for accurate measurements by underwater camera systems," *Sensors*, vol. 15, pp. 30810–30826, 2015.
11. N. Gracias and J. Santos-Victor, "Underwater video mosaics as visual navigation maps," *Comput. Vision Image Understanding*, vol. 79, pp. 66–91, 2000.
12. C. Kunz and H. Singh, "Stereo self-calibration for seafloor mapping using AUVs," in *Proceedings of the IEEE Autonomous Underwater Vehicles*, Monterey, CA, USA, Sep. 2010, pp. 1–7.
13. P. Silvatti, F. A. Salve Dias, P. Cerveri, and R. M. Barros, "Comparison of different camera calibration approaches for underwater applications," *J. Biomech.*, vol. 45, no. 6, pp. 1112–1116, 2012.
14. D. Lodi Rizzini, F. Kallasi, F. Aleotti, F. Oleari, and S. Caselli, "Integration of a stereo vision system into an autonomous underwater vehicle for pipe manipulation tasks," *Comput. Electr. Eng.*, pp. 1–12, 2016.
15. R. Li, C. Tao, W. Zou, R. G. Smith, and T. A. Curran, "An underwater digital photogrammetric system for fishery geomatics," *Int. Arch. Photogramm. Remote Sens.*, vol. 31, pp. 317–323, 1996.
16. M. R. Shortis and E. S. Harvey, "Design and calibration of an underwater stereo-video system for the monitoring of marine fauna populations," *Int. Arch. Photogramm. Remote Sens.*, vol. 32, no. 5, pp. 792–799, 1998.

17. Meline, J. Triboulet, and B. Jouvencel, "A camcorder for 3D underwater reconstruction of archeological objects," in *Proceedings of the OCEANS.*, Seattle, WA, USA, Sep. 2010, pp. 1–9.

18. F. Menna, E. Nocerino, S. Troisi, and F. Remondino, "A photogrammetric approach to survey floating and semi-submerged objects," *Proc. SPIE*, vol. 8791, pp. 14–16, 2013.

19. T. Treibitz, Y. Y. Schechner, and H. Singh. "Flat refractive geometry," in *Proceedings of the IEEE Conference on Computer Vision and Pattern Recognition*, Anchorage, AK, USA, Jun. 2008, pp. 1–8.

20. B. Marshall and D. Eppstein, "Mesh generation and optimal triangulation," in *Computing in Euclidean Geometry*, 2nd ed, Singapore: World Scientific Publishing, 1995, pp. 47–123.

21. Q. T. Luong and O. D. Faugeras, "The fundamental matrix: Theory, algorithms, and stability analysis," *Int. J. Comput. Vision*, vol. 17, no. 1, pp. 43–75, 1996.

22. R. Li, H. Li, W. Zou, R. G. Smith, and T. A. Curran, "Quantitative photogrammetric analysis of digital underwater video imagery," *IEEE J. Oceanic Eng.*, vol.22, no. 2, pp.364–375, 1997.

23. Sedlazeck and R. Koch, "Calibration of housing parameters for underwater stereo-camera rigs," in *Proceedings of the British Machine Vision Conference*, University of Dundee, Aug. 2011, pp. 1–11.

24. Agrawal, S. Ramalingam, Y. Taguchi, and V. Chari, "A theory of multilayer flat refractive geometry," in *Proceedings of the IEEE Conference on Computer Vision and Pattern Recognition*, Providence, RI, USA, Jun. 2012, pp. 3346–3353.

25. X. Chen and Y. H. Yang, "Two view camera housing parameters calibration for multi-layer flat refractive interface," in *Proceedings of the IEEE Conference on Computer Vision and Pattern Recognition*, Columbus, OH, USA, Sep. 2014, pp. 524–531.

26. G. Telem and S. Filin, "Photogrammetric modeling of underwater environments," *ISPRS J. Photogramm. Remote Sens.*, vol. 65, no, 5, pp. 433–444, 2010.

27. T. Dolereit, U. F. von Lukas, and A. Kuijper, "Underwater stereo calibration utilizing virtual object points," in *Proceedings of the OCEANS*, Genoa, Italy, May 2015, pp. 1–7.

28. Anwer, S. S. A. Ali, A. Khan, and F. Meriaudeau, "Underwater 3-D scene reconstruction using Kinect v2 based on physical models for refraction and time of flight correction," *IEEE Access*, vol. 5, pp. 15960–15970, 2017.

29. Jordt-Sedlazeck and R. Koch, "Refractive calibration of underwater cameras," in *Proceedings of the European Conference on Computer Vision*, Firenze, Italy, Oct. 2012, pp. 846–859.

30. Li, L. Heng, K. Kooser, and M. Pollefeys, "A multiple-camera system calibration toolbox using a feature descriptor-based calibration pattern," in *Proceedings of the IEEE International Conference on Intelligent Robots and Systems*, Tokyo, Japan, Jan. 2013, pp. 1301–1307.

31. M. Marcon, A. Sarti, and S. Tubaro, "Multicamera rig calibration by double-sided thick checkerboard," *IET Comput. Vision*, vol .11, no. 6, pp. 448–454, 2017.

32. S. Zhang and P. S. Huang, "Novel method for structured light system calibration," *Opt. Eng.*, vol. 45, no. 8, pp. 1–8, 2006.

33. Z. Wang, W. Wu, X. Xu, and D. Xue, "Recognition and location of the internal corners of planar checkerboard calibration pattern image," *Appl. Math. Comput.*, vol. 185, no. 2, pp. 894–906, 2007.

34. K. Deb, "An introduction to genetic algorithms," *Sadhana*, vol. 24, no. 5, pp. 293–315, 1999.

35. J. Zhang, H. Zhu, C. Yang, Y. Li, and H. Wei, "Multi-objective shape optimization of helico-axial multiphase pump impeller based on NSGA-II and ANN," *Energy Convers. Manage.*, vol. 52, no. 1, pp. 538–546, 2011.

36. K. Deb, A. Pratap, S. Agarwal, and T. Meyarivan, "A fast and elitist multiobjective genetic algorithm: NSGA-II," *IEEE Trans. Evol. Comput.*, vol. 6, no. 2, pp. 182–197, 2002.

37. N. Srinivas and K. Deb, "Muiltiobjective optimization using nondominated sorting in genetic algorithms," *Evol. Comput.*, vol. 1, no. 3, pp. 221–248, 1994.

38. G. Renner and A. Ekart, "Genetic algorithms in computer aided design," *Comput. Aided Des.*, vol. 35, no. 8, pp. 709–726, 2003.

39. K. D. Tran, "Elitist non-dominated sorting GA-II (NSGA-II) as a parameter-less multi-objective genetic algorithm," in *Proceedings of the IEEE Southeast Conference*, Ft. Lauderdale, FL, USA, Apr. 2005, pp. 359–367.

CHAPTER 4

Joint Anchor-Feature Refinement for Real-Time Accurate Object Detection in Images and Videos

4.1 INTRODUCTION

Object detection is one of the fundamental and challenging areas of research in computer vision. With rapid advances in deep learning, convolutional neural networks (CNN) have demonstrated the state-of-the-art performance in this task. Zhao et al. presented an overview of modern object detection approaches [1]. From this review, we can see that two-stage detectors represented by RCNN family [2] and RFCN [3] usually attain an accurate yet slightly slow performance. On the contrary, by detecting objects in a one-step fashion, single-stage detectors [4, 5] are able to run in real time with reasonably modest accuracy. Therefore, fast, accurate detection remains a challenging problem for real-world applications.

It is instructive that the two-stage method induces high accuracy, while the single-stage detector has a desirable inference speed. This inspires us to investigate the reasons. In our opinion, the high accuracy of two-stage approaches comes with two advantages: (1) Two-step regression and (2) relatively accurate features for detection. In detail, two-stage detectors firstly regress predefined anchors with the aid of region proposal [2], and this operation significantly eases the difficulty of final localization. Besides, an RoI-wise subnetwork [2] is appended to the region proposal part, so features in the region of interest can be leveraged for final detection. By contrast, there are two drawbacks in the single-stage paradigm: (1) Detection head directly regresses coordinates from predefined anchors, but most anchors are far from matching object regions, and (2) classification information comes from probably inaccurate locations, where features could not be precise enough to describe objects. Referring to Figure 4.1a, it is relatively difficult to regress predefined anchors to precisely surround the object (e.g., the dog in Figure 4.1). Moreover, as feature sampling locations follow predefined anchor regions, detection features for small-scale anchors cannot cover the entire object region, while that for large-scale anchors weaken the object because of background. On the contrary, owing to region proposal, the two-stage methods detect the dog using a better initialization (see Figure 4.1b). Thus, the strengths of two-stage methods exactly reflect the single-stage drawbacks that lead to relatively lower detection accuracy. Although Zhang et al. developed RefineDet [6] to introduce two-step regression to the single-stage detector, it still failed to capture accurate detection features. That is, predefined

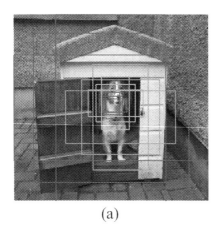 

(a)   (b)

FIGURE 4.1 Comparison of single-stage anchors and RPN outputs. For better visualization, only several key boxes are demonstrated. (a) Multi-scale SSD anchors. (b) RPN outputs in Faster RCNN.

feature sampling locations are not precise enough for describing refined anchor regions. (Note that detailed comparison between RefineDet and our approach will be presented in Section 4.3.2.) Thus, there is an imperative need for further overcoming these single-stage limitations for real-time accurate object detection.

In addition, most researches have largely focused on detecting object statically, ignoring temporal coherence in real-world applications. Detection in real-world scenes was introduced by ImageNet video detection (VID) dataset [7]. To the best of our knowledge, main ideas of temporal detection include (1) post-processing [8], (2) tracking-based location [9, 10], (3) feature aggregation with motion information [9, 11–14], (4) RNN-based feature propagation [13, 15–17], and (5) batch-frame processing (i.e., tubelet proposal) [18]. All these ideas are attractive in that they are able to leverage temporal information for detection, but they also have respective limitations. In brief, methods (1)–(4) borrow other tools (e.g., tracker, optical flow, and LSTM) for temporal analysis. Methods (3) and (4) focus on constructing superior temporal features. Nevertheless, they detect objects following the static mode. Method (5) works in a non-causal offline mode that prohibits this approach from real-world tasks. Furthermore, most recent works pay excessive attention to accuracy so that high computational costs could affect time efficiency. Thus, a novel temporal detection mode should be developed for videos.

Overcoming aforementioned single-stage drawbacks, a dual refinement mechanism is proposed in this chapter for static and temporal visual detection, namely, anchor-offset detection. This joint anchor-feature refinement includes an anchor refinement, a feature location refinement, and a deformable detection head. The anchor refinement is developed for two-step regression, while the feature location refinement is proposed to capture accurate single-stage features. Besides, a deformable detection head is designed to leverage this dual refinement information. Based on the anchor-offset detection, we propose three approaches for object detection in images and videos. Firstly, a dual refinement network (DRNet) is proposed. DRNet is designed for static detection, where a multi-deformable head is developed for diversifying detection receptive fields for more contextual information. Secondly, temporal refinement networks (TRNet) are designed, which perform anchor refinement across time for video detection. Thirdly, temporal dual refinement networks (TDRNet) are developed that extend the anchor-offset detection toward temporal tasks.

Additionally, for temporal detection task, we propose a soft refinement strategy to match object motion with previous refinement information. Our proposed DRNet, TRNet, and TDRNet are validated on PASCAL VOC [19], COCO [20], and ImageNet VID [7] datasets. As a result, our methods achieve a real-time inference speed and considerably improved detection accuracy. Contributions are summarized as follows:

- Starting with drawbacks of single-stage detectors, an anchor-offset detection is proposed to perform two-step regression and capture accurate object features. The anchor-offset detection includes an anchor refinement, a feature-offset refinement, and a deformable detection head. Academically, without region-level processing, this joint anchor-feature refinement achieves single-stage region proposal. Thus, the anchor-offset detection bridges single-stage and two-stage detection so that it is able to induce a new detection mode.

- A DRNet based on the anchor-offset detection and a multi-deformable head is developed to elevate static detection accuracy while maintaining real-time inference speed for the image detection task.

- As a new temporal detection mode for video detection task, a TRNet and a TDRNet are proposed based on the anchor-offset detection without the aid of any other temporal modules. They are characterized by better accuracy vs. speed trade-off and have a concise training process without the requirement of sequential data. In addition, a soft refinement strategy is designed to enhance the effectiveness of refinement information across time.

- The single-stage DRNet maintains fast speed while acquiring significant improvements in accuracy, i.e., 84.4% mean average precision (mAP) on VOC 2007 test set, 83.6% mAP on VOC 2012 test set, and 42.4% AP on COCO test-dev. Based on VID 2017 validation set, DRNet sees 69.4% mAP, TRNet achieves 66.5% mAP, and TDRNet obtains 67.3% mAP.

The remainder of this chapter is organized as follows. Section 4.2 presents the related works. Including anchor-offset detection and multi-deformable head, DRNet is elaborated in Section 4.3. Section 4.4 presents TRNet and TDRNet in detail, and Section 4.5 provides the experimental results and discussion. Conclusions are summarized in Section 4.6.

4.2 REVIEW OF DEEP LEARNING-BASED OBJECT DETECTION

4.2.1 CNN-Based Static Object Detection

Deep learning methods have recently dominated the field of object detection [1]. Two-stage detectors [2, 3, 21, 22] usually detect objects by region proposal, location, and classification. For example, inspired by Faster RCNN [2] and RFCN [3], CoupleNet [21] leveraged both region-level and part-level features to express a variety of challenging object situations, which achieved considerable detection accuracy but it only ran at 8.2 FPS. As groundbreaking works, YOLO [4] and SSD [5] localized and classified objects using a single-shot network for real-time detection. Recently, many revised single-stage versions have emerged [6, 23–28]. Typically, in favor of small object detection, Lin et al. developed a RetinaNet to formulate the single-shot network as an FPN [29] fashion for propagating information in a top-down manner to enlarge shallow layers' receptive field [23]. Redmon and Farhadi proposed YOLOv3 with DarkNet53 and multi-scale anchor for fast, accurate detection [28]. Zhang et al. designed a RefineDet to introduce two-step regression to single-stage pipeline [6]. RefineDet adjusted predefined anchors for more precise localization. However, its detection features were still fixed on predefined positions, failing to precisely describe refined anchor regions. In short, although single-stage methods have a superiority in speed, two-stage methods still dominate detection accuracy on generic benchmarks [7, 19, 20]. Hence, we are motivated to analyze single-stage drawbacks from two-stage merits (analyzed in Section 4.1) and construct DRNet with both competitive accuracy and fast speed.

4.2.2 Temporal Object Detection

To detect objects in temporal vision, some post-processing methods have been first investigated to merge multi-frame results, and then tracker-based detection, motion-guided feature aggregation, RNN-based feature integration, and tubelet proposal are studied by the research community. Han et al. proposed a SeqNMS to discard temporally interrupted bounding boxes in the non-maximum suppression (NMS) phrase [8]; Feichtenhofer et al. combined RFCN and a correlation-filter-based tracker to boost recall rate [10]. Based on motion estimation with optical flow, Zhu et al. devised a temporally adaptive key frame, scheduling to effective feature aggregation [12]; Chen et al. and Liu et al. took advantage of long short-term memory to propagate CNN features across time [16, 17].

However, the temporal analysis capacity in the abovementioned methods is obtained from other temporal tools. Although some methods focused on how to construct superior temporal features, they still remained inapposite static detection mode. As a typical offline detection mode, Kang et al. reported a TPN for tubelet proposal (i.e., temporally propagated boxes) so that multiple frames could be simultaneously processed to improve temporal consistency [18]. However, this batch-frame mode struggled to be qualified for real-world tasks. On the contrary, without the aid of any other temporal tools, we developed a novel real-time online detection mode for videos using the idea of refinement. That is, refined anchors and refined feature sampling locations are generated with key frames, which would be temporally propagated for detection. Compared to most video detectors, our methods have a concise training process without the need for sequential images.

4.2.3 Sampling for Object Detection

It is widely accepted that spatial sampling is important to construct robust features. For example, Peng et al. detected objects by an improved multistage particle window that can sample a small number of key features for detection [30]. In terms of CNN, canonical convolution is based on a square kernel that is not suited enough to variform objects. For augmenting the spatial sampling locations, Dai et al. proposed deformable convolutional networks to combat fixed geometric structures in traditional convolution operation. The deformable convolution significantly boosted the detection accuracy of RFCN [31]. As for video detection, Bertasius et al. used the deformable convolutions across time and constructed robust features for temporally describing objects [14]. Zhang et al. designed a feature consistency module with deformable convolution to reduce inconsistency in the single-stage pipeline [32]. Wang et al. proposed guided anchoring for RPN, Faster RCNN, and RetinaNet to achieve a higher quality region proposal [33]. Creatively, we tend to capture accurate single-stage features, and more specifically, refined feature locations are produced based on refined anchors. Moreover, we propagate refinement information across time for video detection.

4.3 DUAL REFINEMENT NETWORK

In this section, the proposed DRNet will be presented. The network architecture is first briefed, and then, we will demonstrate how to overcome two key single-stage drawbacks with anchor-offset detection. Next, our designed multi-deformable head is delineated, followed by the training and inference.

4.3.1 Overall Architecture

As shown in Figure 4.2, our proposed architecture is a single-shot network with a forward backbone for feature extraction. The network generates a fixed number of bounding boxes and corresponding classification scores, followed by the NMS for duplicate removal. Inheriting from RefineDet [6], there is an anchor refinement module (ARM) and an object detection module (ODM) for two-step regression. ARM regresses coordinates for refined anchors, and then feature offsets are predicted using anchor offsets. In ODM, a creative detection head is designed with deformable convolution for final classification and regression, whose inputs are ODM features, refined anchors, and feature offsets. Furthermore, a multi-deformable head is developed with multiple detection paths to leverage contextual information for detection.

4.3.2 Anchor-Offset Detection

4.3.2.1 From SSD to RefineDet, then to DRNet

As illustrated in Figure 4.3a, SSD directly detects objects with ARM features, whereas RefineDet adopts FPN for strong semantic information. Moreover, RefineDet develops an anchor refinement for more precision localization and a negative anchor filtering for addressing extreme class imbalance problem. In our DRNet, we inherit anchor refinement but discard the negative anchor filtering since training with hard negative mining [5] induces a similar effect. More specifically, a feature location refinement and a deformable detection head are proposed to combat another key drawback in the single-stage paradigm, i.e., inaccurate feature sampling locations.

In general, detection in traditional SSD-like manner is based on handcrafted anchors which are rigid and usually inaccurate. Predefined anchors and fixed feature locations could not be suited enough to regress and classify objects (see the left top in Figure 4.3b). Through preliminary localization, refined anchors in RefineDet are in favor of more precise coordinate prediction. However, RefineDet still uses inaccurate feature sampling locations (see the right top in Figure 4.3b) for regression and classification. That is, the anchor refinement would incur serious anchor-feature misalignment. Thus, it is defective that the anchor refinement is

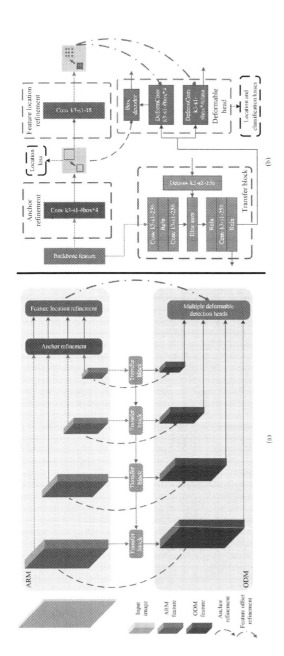

FIGURE 4.2   The schematic layout of the proposed DRNet. Refined anchors are produced by coarse regression with ARM features, and they are first employed to predict feature offsets, namely, feature location refinement. The detection head utilizes ODM feature maps, refined anchors, and refined feature sampling locations to detect objects, i.e., anchor-offset detection. A multi-deformable head is designed for rich contextual information. (a) Overall framework. Each anchor refinement induces a feature location refinement. (b) Design details. Convolution is detailed with kernel size, stride, and output channel size. Only one detection path with 3×3 convolutional kernel is shown.

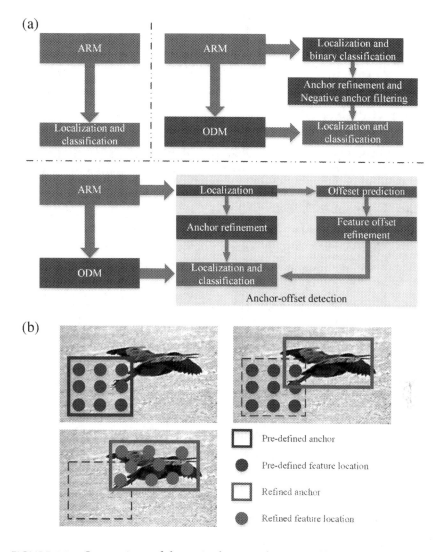

FIGURE 4.3 Comparison of three single-stage detectors. (a) Structure sketch of SSD (left, top), RefineDet (right, top), and DRNet (bottom). (b) Detection modes of SSD (left, top), RefineDet (right, top), and DRNet (left, bottom). (b) shows the main idea of the anchor-offset detection.

leveraged alone. Overcoming these difficulties, our designed anchor-offset detection is able to achieve two-step regression and capture more accurate detection features in a single-stage pipeline (see the left bottom in Figure 4.3b). This joint anchor-feature refinement manner is more reasonable than RefineDet.

4.3.2.2 Anchor Refinement

This process is analogous to RefineDet, i.e., ARM generates refined anchors that provide better initialization for the second-step regression. A location head performs convolution to generate anchor offset ar using backbone-based ARM features f_{ARM}. That is, $ar = W_{ar} * f_{ARM}$, where $*$ denotes convolution (W is the learnable convolutional weight). Note that ar is the coordinate offset from original anchors. Anchor refinement is urgently necessary. On the one hand, it highly relieves the difficulty of localization. On the other hand, it can guide feature location refinement.

4.3.2.3 Deformable Detection Head

According to deformable convolutions [31], a deformable detection head is designed to leverage the refinement information. The standard detection head in SSD uses a regular 3×3 grid $\mathcal{R}$ to predict category probability and coordinates for a feature map cell. In the meantime, through careful anchor design, the respective field of $\mathcal{R}$ can describe a specific anchor region. Thus, the prediction can be given as $P_{p_0} = \sum_{p \in \mathcal{R}} w(p) \cdot f_{ODM}(p)$, where P is the prediction of category probability or coordinate offset; w is the convolution weight; p represents positions in $\mathcal{R}$ while p_0 is the center; f_{ODM} denotes ODM features.

However, the respective field of $\mathcal{R}$ usually fails to describe the refined anchor region (see the right top of Figure 4.3b). Thereby, allowing $\mathcal{R}$ to deform to fit various anchor changes, the deformable detection head is developed to capture accurate features with the feature offset δp,

$$P_{p_0} = \sum_{p \in \mathcal{R}} w(p) \cdot f_{ODM}(p + \delta p). \tag{4.1}$$

4.3.2.4 Feature Location Refinement

The offset set $\Delta p = \{\delta p\}$ is computed with the input feature in the original deform pipeline,

$$\Delta p = W_{fr} * f_{ODM}, \tag{4.2}$$

where W_{fr} is the convolutional weight. Nevertheless, there is a strong demand for describing the refined anchor regions with the deformed grids. Therefore, our feature offsets are predicted based on anchor offsets, i.e., feature location refinement,

$$\Delta p = W_{fr} * ar. \tag{4.3}$$

In detail, this operation is a convolution with 1×1 kernel. Since each spatial element in ar is coordinate offsets for refined anchors, its channel information is fused for feature location refinement. Note that anchor offsets and feature offsets are different in tensor shape. A man-made function can be designed to map anchor offsets to feature offsets, but we adopt a learnable mapping. Although the man-made manner can also promote refined feature locations to describe refined anchors, it still generates a conventional regular feature sampling region. Thus, the feature location refinement adopts a learnable manner to produce flexible feature sampling locations from multiple anchor offsets.

In this way, the refined feature locations can describe refined anchor regions more effectively. We call this detection mode anchor-offset detection, which can be formulated as

$$P_{local} = (W_{local} * (f_{ODM}, \Delta p)) \oplus (ar \oplus ao)$$
$$P_{conf} = W_{conf} * (f_{ODM}, \Delta p). \tag{4.4}$$

where $\oplus$, ao represent anchor decoding operation [132] and the original anchor, respectively; $W * (f, \Delta p)$ denotes deformable convolution with W as the weight. As ar is the coordinate offset from ao, $ar \oplus ao$ is the refined anchor. The operation of two $\oplus$ is two-step regression that elevates the precision of localization, while Δp is the feature offset that constructs the accurate single-stage detection features.

4.3.3 Multi-deformable Head

CoupleNet developed local and global FCN to detect objects [21]. The local FCN focused local features in a region proposal while the global one paid attention to the whole region-level features. In this way, more semantic information and underlying object relation are exploited for high-quality detection. Thus, taking aim at describing the object using original, shrunken, and expansile region-level features, a multi-deformable head is developed for the single-stage detector. The shrunken region-level features are in favor of leveraging local messages while the expansile region-level features contain more contextual information and object relation.

In this way, multiple detection head is designed with different respective field sizes, inducing multiple detection paths. As shown in Figure 4.4, each of the detection path is an anchor-offset detection, and their feature location refinement is independent. Besides, their results are fused with an element-wise summation.

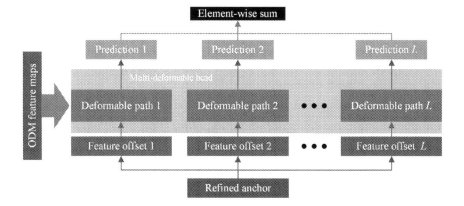

FIGURE 4.4 Multi-deformable head. It is designed with different detection respective field. Multiple detection paths are induced, where feature location refinement is independent for each path. Their results are fused by summation.

The detection based on L deformable paths can be given as

$$P_{local} = \left(\sum_{l=1}^{L} W_{local_l} * (f_{ODM}, \Delta p_l) \right) \oplus (ar \oplus ao)$$

$$(4.5)$$

$$P_{conf} = \sum_{l=1}^{L} W_{conf_l} * (f_{ODM}, \Delta p_l).$$

4.3.4 Training and Inference

As for predefined anchor setting, each feature layer is associated with one specific scale of anchors, i.e., the anchor size of [32, 64, 128, 256] is adopted for four-scale feature maps from low level to high level, and three anchors are tiled at each feature map cell with aspect ratios of [1.0,2.0,0.5]. In terms of optimization, an SGD optimizer with 0.9 momentum and 0.0005 weight decay is employed to train the whole network. Because of different sizes of datasets, the learning rate schedule is diverse for each dataset, which will be briefed later.

A multitask objective is designed to train DRNet including two localization losses, $\mathcal{L}_{loc-ARM}, \mathcal{L}_{loc-ODM}$, and a confidence loss, $\mathcal{L}_{conf}$, i.e.,

$$\mathcal{L} = \frac{1}{N_{ARM}} \mathcal{L}_{loc-ARM} + \frac{1}{N_{ODM}} (\mathcal{L}_{loc-ODM} + \mathcal{L}_{conf}), \text{ where } N \text{ is the number of}$$

positive boxes in ARM and ODM. $\mathcal{L}_{loc} = \sum_{i=1}^{N} sommoth\ L1(p_i - g_i^*)$, where

g_i^* is the ground truth coordinates of the ith positive anchor. Before the computation of $\mathcal{L}_{loc}$, anchors should be determined to be positive or negative based on jaccard overlap [5]. We handle original anchors and refined anchors for $\mathcal{L}_{loc-ARM}$ and $\mathcal{L}_{loc-ODM}$, respectively, by the following processes. Firstly, each ground truth box is matched to anchors with the best jaccard overlap, then anchors with >0.5 overlap will be matched to corresponding ground truth box. Let c_i^{cls} be the probability that the ith predicted box belongs to class cls ($cls = 0$ for background). $\mathcal{L}_{conf} = -\sum_{i=1}^{N} \log(c_i^{cls}) - \sum_{i=1}^{\delta N} \log(c_k^0)$, where δN negative anchors are selected by hard negative mining [5]. This operation selects a part of negative boxes with top loss values for training to address the problem with extreme foreground-background class imbalance, and $\delta = 3$.

In the inference phase, DRNet predicts confident object candidates (confident scores >0.01) in the manner of anchor-offset detection and multi-deformable head. Subsequently, these candidates are processed by NMS with 0.45 jaccard overlap pre class and retain top 200 (for COCO) or 300 (for VOC and VID) high confident objects as the final detections.

4.4 TEMPORAL DUAL REFINEMENT NETWORKS

In this section, we present how to propagate refined anchors and refined feature sampling locations across time. Next, TRNet and TDRNet are formed. Then, we also describe the proposed soft refinement strategy.

4.4.1 Architecture

As shown in Figure 4.5a, a reference generator (RG) and a refinement detector (RD) are designed in this section, both of which are constructed with similar structure, i.e., canonical SSD framework [5] with 4-scale detection features. However, RG and RD have different training mode, parameters, and outputs. Like ARM in DRNet, RG predicts refinement information including refined anchors or both refine anchors and feature offsets. Similar to ODM in DRNet, RD takes over RG's outputs as references and detects objects frame by frame. If RG only predicts refined anchors, the framework is called TRNet. When feature offsets are also

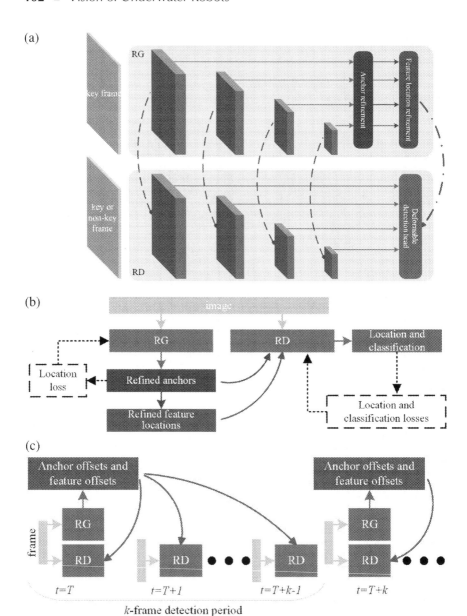

FIGURE 4.5   Designs of the proposed TDRNet. (a) Network structure for RG and RD. RG generates refinement information while RD performs final detection. RG's outputs (i.e., anchor offsets and feature location offsets) serve as RD's inputs. (b) The training phase. (c) The testing phase.

predicted by RG, the anchor-offset detection with a deformable detection head is also employed by RD, and we call this structure TDRNet. That is, compared to TDRNet (see Figure 4.5a), TRNet does not contain the module of feature location refinement, and its detection head is composed of traditional convolutions.

4.4.2 Training

In general, temporal detectors usually have a complex training process with sequential images. For example, TSSD developed a multistep training strategy [17], and the initialization for multi-frame regression layer in TPN is complicated [18]. Conversely, the training process for TRNet and TDRNet is refreshingly concise, and we also eliminate the need for sequential training images. As shown in Figure 4.5b, during the training process, RG and RD play similar roles to DRNet's ARM and ODM, respectively. Thereby, both RG and RD can be trained with static images following DRNet's basic training settings and loss functions.

4.4.3 Inference

Consider a video as an image sequence, i.e., $V = \{I_0, I_1, \ldots, I_M\}$. TRNet and TDRNet attempt to obtain frame-level detections $\{D_0, D_1, \ldots, D_M\}$, where D_m contains the boxes and class predictions of I_m. RG takes over I_m and outputs anchor offset ar and feature offset Δp,

$$ar_m, \Delta p_m = RG(I_m). \qquad (4.6)$$

Then, RD detects objects with I_m, ar, and Δp,

$$D_m = RD(I_m, ar, \Delta p)$$

$$= \begin{cases} (W_{local} * (f_{I_m}, \Delta p)) \oplus (ar \oplus ao) \\ W_{conf} * (f_{I_m}, \Delta p), \end{cases} \qquad (4.7)$$

where f_{I_m} is the feature extracted from I_m.

Despite the similar detection manner, it is apparent that RD is more computationally efficient than DRNet. Therefore, considering the temporal context in temporal vision, a key frame duration is used for RG to pursue a better trade-off between accuracy and speed. That is, only key frames will be processed by RG, while non-key frames are detected by RD with previous RG's outputs. Mathematically, in equation (4.7), $ar = ar_m, \Delta p = \Delta p_m$ for key frames, whereas ar, Δp are from the previous key frame when detecting non-key frames. In this manner, ar and Δp are propagated as the temporal information. As illustrated in Figure 4.5c, using the first frame in a period, RG generates refinement references that will survive k time stamps. Then, RD detects objects based on these references for the whole period. It is apparent that frequent reference update would lead to higher detection accuracy and more computational costs, so the trade-off between accuracy and speed can be adjusted by different k settings.

Taking aim at adapting to various object motion, a soft refinement strategy is proposed with a soft coefficient e. In SSD, the intent of designing anchor is to use numerous boxes to cover the whole image as the prior knowledge, but significantly discarding anchor diversity, the refined anchors tend to surround foreground. Refined anchors are in favor of static detection, but objects in videos have a variety of motion properties or pose changes. Hence, the soft refinement strategy is designed to retain the anchor diversity for relatively long temporal detection period. The soft refinement can be given as $ar_s = ar \times e$, where ar_s is the soft anchor offset, and as a scalar, $e \in [0,1]$ multiplies each element in a tensor. Because ar is the offset from original anchors, $ar \times e$ can relax the intensity of anchor refinement by reducing offset magnitude so that refined anchors can be loosely scattered around objects. Referring to Figure 4.6, refined anchors are computed by the first frame in the period, then some key refined anchors are visualized in the second, fourth, and eighth frames. Without the soft refinement strategy, they gradually fail to be precisely aware of objects across time. For example, when $e = 1$, refined anchors cannot surround the head of a sheep in the eighth frame. This phenomenon causes regression difficulties for RD, prohibiting k from increasing. That is, $e = 1$ incurs that the refinement information can hardly be propagated in a relatively long range of time series. When $e = 0.75$ or 0.5, this drawback is mitigated so that the detection period can be longer for a better trade-off between accuracy and speed.

4.5 EXPERIMENTS AND DISCUSSION

Our methods are implemented under the Pytorch framework. The training and experiments are carried out on a workstation with an Intel 2.20 GHz Xeon(R) E5-2630 CPU, NVIDIA TITAN-1080 GPUs, CUDA 8.0, and cuDNN v7. Our approaches are trained and evaluated on PASCAL VOC [19], COCO [20], and ImageNet VID [7] datasets. Furthermore, we applied TDRNet to online underwater object detection.

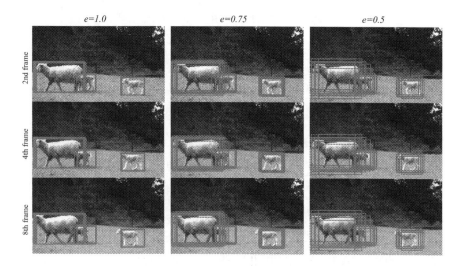

FIGURE 4.6   Soft refinement strategy. To match object motion with the previous refinement, a soft coefficient e is introduced. Computed with the first frame, key refined anchors are demonstrated in the second, fourth, and eighth frames. The decrease of e reduces refinement intensity, producing loosely scattered refined anchors. That is, it is seen that refined anchors based on $e = 1$ are compact. In contrast, refined anchors based on $e = 0.75$ are more loosely scattered than that based on $e = 1$, while refined anchors based on $e = 0.5$ are more loosely scattered than that based on $e = 0.75$.

4.5.1 Ablation Studies of DRNet320-VGG16 on VOC 2007

Experiments on PASCAL VOC 2007 are first conducted with VGG16 [34] as the backbone to study the proposed dual refinement mechanism in detail. In this section, the models are trained on the union set of VOC 2007 trainval and VOC 2012 trainval (16,551 images, denoted as "07 + 12") and evaluated on VOC 2007 test set (4,952 images). We use mAP to describe the detection accuracy. For the convenience of comparison, RefineDet without negative anchor filtering is employed as the baseline, whose mAP is 79.1% in our reproduced Pytorch implementation. (Note that it is 79.5% in original Caffe implementation.) The changes of mAP caused by various model designs are shown in Table 4.1.

4.5.1.1 Anchor-Offset Detection

The anchor-offset detection contains an anchor refinement, a feature location refinement, and a deformable detection head, the first of which has been studied by [6], so we focus on the latter two components. At first, the

TABLE 4.1   Ablation studies of DRNet320 on VOC 2007. The baseline is 79.1%

Component	DRNet320-VGG16					
Multi-deformable head?			√			√
Feature location refinement?		√	√			√
Deformable detection head?	√	√	√		√	√
BN?						
mAP(%)	78.3	79.8	80.5	81.1	81.7	82.0

deformable detection head without feature location refinement is tested. Following [31], the offsets are computed with ODM features (referring to equation (4.2)). As a result, this change leads to a 0.8% mAP drop. In our opinion, this should be attributed to improper offsets. That is, the refined anchors are computed with ARM, while the feature offsets are from ODM so they are independent, making refined feature locations still fail to describe refined anchor regions.

The refined anchors have been displayed in Figure 4.6, so the refined feature sampling locations are also demonstrated in Figure 4.7 to better explain the advantages of the proposed anchor-offset detection. For better visualization, only the sampling centers (i.e., the center dot in the left bottom Figure 4.3b) are demonstrated. Referring to dots in the first row Figure 4.7, the predefined detection features are regularly fixed on feature maps (their locations are mapped to the original images for visualization). This design is justified for the traditional SSD since anchors are also tiled

FIGURE 4.7   Visualization of refined feature sampling locations for *Conv5_3*. For better visualization, only the sampling centers (i.e., the center dot in the left bottom of Figure 4.3b) are demonstrated. The original sampling centers are illustrated with dots in the first row, which are regularly tiled on images. The dots in the second row show the refined sampling centers that have a stronger capability of describing objects. These images are from VOC, COCO, and ImageNet VID.

in the same manner. However, the refined anchors tend to surround objects for more precision localization (see Figure 4.6), so it is reasonable that the feature locations should have the same tendency. As shown with dots in the second row, gathering toward objects, the refined feature locations are more suitable for regression and classification. Moreover, in some areas away from objects, the refined feature locations would not blindly shift toward targets so that the detection capability for the whole image can be maintained.

Therefore, the operation of the proposed feature location refinement is crucial to capture accurate detection features. Following the pipeline of anchor-offset detection, the refined feature locations are tightly associated with refined anchors. Thus, a 0.7% mAP rise (i.e., 79.8% vs. 79.1%) is induced.

4.5.1.2 Multi-deformable Head

For leveraging more contextual information for detection, multiple detection paths are devised with various respective field sizes, or convolution kernel size and dilation. The effectiveness of various multi-deformable designs is shown in Table 4.2. At first, the 1×1 grid is employed to utilize shrunken region-level features, but it incurs negligible effectiveness. The 1×1 grid should have focused on most suitable local parts for detection, but feature offsets are computed with anchor offsets in our pipeline, ignoring suitable local parts. Then, the 3×3 grid with dilation is devised as one of the detection paths, but it leads to a 0.4% drop in mAP. Although it expands the respective field, the dilated 3×3 grid splits features and fails to describe objects effectively. Covering this shortage, the 5×5 grid without dilation works more effectively, and it invites a 0.7% mAP rise (i.e., 80.5% vs. 79.8%) since more contextual information is involved. Moreover, the 1×1 detection path is removed, and this more efficient design still can reach 80.3% in mAP. These comparisons also indicate that the improvement comes from above-analyzed reasons rather than increasing parameter size.

TABLE 4.2 Effectiveness of various multi-deformable head designs. A variety of detection paths with different convolutional kernel size (*ks*) and dilation (*di*) are used to validate the efficacy our designs

$ks = 5 \times 5, di = 1$?			√	
$ks = 3 \times 3, di = 2$?		√		
$ks = 1 \times 1, di = 1$?	√	√	√	
$ks = 3 \times 3, di = 1$?	√	√	√	√
mAP(%)	79.8	79.4	80.5	80.3

4.5.1.3 Toward More Effective Training

Batch normalization [35] is introduced to the backbone for more effective training, and a significant improvement in accuracy is incurred, i.e., 81.1% mAP. Then, the anchor-offset detection and the multi-deformable head further boost the performance. Referring to Table 4.1, removing the multi-deformable head leads to a 0.3% drop in mAP, and removing the anchor-offset detection invites another 0.6% mAP drop. Thus, our designs are still efficient, making a superior detection performance with such a small input image, i.e., 82.0% mAP and 320×320 input size.

4.5.2 Results on VOC 2007

We use the initial learning rate of 0.001 for the first 130 training epochs, then use the learning rate of 0.0001 for the next 40 epochs and 0.00001 for another 40 epochs. Referring to Table 4.3, our DRNet320 achieves 82.0% mAP, surpassing all methods with such small inputs by a large margin. When compared to SSD300, our method outperforms it by 4.8 points (i.e., 82.0% vs. 77.2%), and DRNet320 further improves mAP by 2.0% as for RefineDet320 (i.e., 82.0% vs. 80.0%). Compared to RFBNet300, our DRNet320 also has 1.5-point higher mAP (i.e., 82.0% vs. 80.5%).

For 512×512 input size, DRNet512 obtains 82.8% mAP that is also competitive with all compared methods. Only Attention CoupleNet [22] has slightly higher mAP than ours (i.e., 82.8% vs. 83.1%). However, Attention CoupleNet uses ResNet101 [36] as its backbone, and its results come with larger input size (i.e., 1,000×600). Besides, Attention CoupleNet introduces extra segmentation annotations to its multi-scale training processing. In addition, DRNet512's inference speed surpasses that of Attention CoupleNet by a large margin (i.e., 32.2 vs. 6.9 FPS). Therefore, the proposed DRNet achieves a better trade-off between accuracy and speed. To relieve the impact of relatively small input size, we leverage multi-scale strategy for testing, and DRNet320 and DRNet512 can obtain 83.9% and 84.4% mAP, respectively.

4.5.3 Results on VOC 2012

More challenging VOC 2012 dataset is employed to evaluate our proposed designs, and we use the union set of VOC 2007 and VOC 2012 trainval sets plus VOC 2007 test set (21,503 images) for training in this experiment and test models on VOC 2012 test set (10,991 images). The learning rate schedule is consistent with VOC 2007 training. Referring

TABLE 4.3 Results on Pascal VOC 2007 and 2012 test dataset. "Train data" is used for VOC 2007 training, and that of VOC 2012 contains an extra VOC 2007 test set. "S" denotes instance segmentation labels. "+" indicates multi-scale testing. Bold fonts indicate the best results

Method	Backbone	Train data	Input size	#Boxes	FPS	mAP	
						VOC 2007	VOC 2012
Two-stage							
Faster RCNN [2]	VGG16	07+12	1,000×600	300	7	73.2	70.4
Faster RCNN [2]	ResNet101	07+12	1,000×600	300	2.4	76.4	73.8
RFCN [3]	ResNet101	07+12	1,000×600	300	9	80.5	77.6
CoupleNet [21]	ResNet101	07+12	1,000×600	300	8.2	82.7	80.4
Attention CoupleNet [22]	ResNet101	07+12+S	1,000×600	300	6.9	83.1	81.0
Single-stage							
YOLO [4]	GoogleNet [39]	07+12	448×448	98	45.0	63.4	57.9
SSD321 [25]	ResNet101	07+12	321×321	17,080	11.2	77.1	75.4
SSD300 [5]	VGG16	07+12	300×300	8,732	120.0*	77.2	75.8
DSOD300 [28]	DenseNet [40]	07+12	300×300	8,732	17.4	77.7	76.3
YOLOv2 [26]	Darknet19	07+12	544×544	1,445	40.0	78.6	73.4
DSSD321 [25]	ResNet101	07+12	321×321	17,080	9.5	78.6	76.3
SSD512 [5]	VGG16	07+12	512×512	24,564	34.7	79.8	78.5
RefineDet320 [6]	VGG16	07+12	320×320	6,375	60.0*	80.0	78.1
RFBNet300 [27]	VGG16	07+12	300×300	8,808	83.0	80.5	–
SSD513 [25]	ResNet101	07+12	513×513	43,688	6.8	80.6	79.4
DSSD513 [25]	ResNet101	07+12	513×513	43,688	5.5	81.5	80.0
RefineDet512 [6]	VGG16	07+12	512×512	16,320	24.1	81.8	80.1
RFBNet512 [27]	VGG16	07+12	512×512	24,692	38.0	82.2	–
DRNet320	VGG16	07+12	320×320	6,375	55.2	82.0	79.3
DRNet512	VGG16	07+12	512×512	16,320	32.2	82.8	80.6
DRNet320+	VGG16	07+12	320×320	6,375	–	83.9	83.1
DRNet512+	VGG16	07+12	512×512	16,320	–	**84.4**	**83.6**

to Table 4.3, our DRNet320 obtains 79.3% mAP that outmatches all compared methods with similar small input sizes. With 512×512 input size, DRNet512 improves the mAP to 80.6%, which validates the effectiveness of our designs once again. Additionally, with multi-scale testing, 83.1% and 83.6% mAP are induced by DRNet320 and DRNet512.

4.5.4 Results on COCO

We perform a thorough analysis on COCO detection dataset, which contains 80 class labels. As in previous work, we also use the union of training images and a subset of validation images (118,278 images, denoted as "trainval35k") for training and test models on test-dev set (20,288 images). The whole network is trained for 70 epochs with a learning rate of 0.001, then for 30 epochs with a learning rate of 0.0001 and another 30 epochs with a learning rate of 0.00001. The main COCO metric denotes as AP, which evaluates detection results at IoU∈ [0.5:0.05:0.95]. AP@0.5, AP@0.75, AP_S, AP_M, and AP_L are also used for deep comparison.

As shown in Table 4.4, some anchor-free methods recently achieve high AP on COCO, which leverage key-point technology [37–39] or full convolution [40] for object detection. DRNet320 achieves the results of 30.5%, which is better than contemporary methods (e.g., RefineDet320, RFBNet300), so our approach can effectively cope with a variety of complex situations with small input resolution. Furthermore, DRNet512 obtains a more competitive AP of 34.3%. Because they have similar AP results, we draw readers' attention to a deep comparison among methods in boldface. (Note that YOLOv3 is not written in bold because its input size is 608.) At first, DRNet512 has huge improvements as opposed to RefineDet512 on all criteria, where our designs are proved to be successful. Moreover, our DRNet512 has the best VOC-like AP@0.5 (i.e., 57.1%) and AP_S (i.e., 17.6%), so our method is more adept at small object detection, owing to the proposed dual refinement mechanism. However, our results on AP@0.75 and AP_L are not comparable with that of some methods. This is caused by two reasons: (1) Two-stage methods use larger input size; (2) ResNet101 or RFB block [26] provides larger effective receptive field for describing large objects [41]. Therefore, our methods are also tested with ResNet101 as the backbone. As a result, more competitive results are induced, i.e., DRNet320ResNet101 delivers 33.5% AP and DRNet512-ResNet101 achieves 38.6% AP.

If multi-scale testing is employed, we see 42.4% AP from DRNet512-ResNet101. Additionally, using MobileNet [42] as the backbone, our

TABLE 4.4 Results on COCO test-dev

Method	Backbone	Train data	AP	AP@0.5	AP@0.75	AP_S	AP_M	AP_L	Time
Anchor-free									
ExtremeNet [31]	Hourglass104	trainval35k	40.2	55.5	43.2	20.4	43.2	53.1	–
CornerNet [30]	Hourglass104	trainval35k	40.6	56.4	43.2	19.1	42.8	54.3	–
FCOS [33]	ResNet101	trainval35k	41.5	60.7	45.0	24.4	44.8	51.6	–
CenterNet [32]	Hourglass104	trainval35k	42.1	61.1	45.9	24.1	45.5	52.8	–
Two-stage									
Faster RCNN† [2]	MobileNet	trainval32k	19.8	–	–	–	–	–	–
Faster RCNN [2]	VGG16	train	24.2	45.3	23.5	7.7	26.4	37.1	147 ms
Faster RCNN [2]	ResNet101	trainval	29.4	48.0	–	9.0	30.5	47.1	–
RFCN [3]	ResNet101	trainval	29.9	51.9	–	10.8	32.8	45.0	110 ms
Deformable Faster RCNN [35]	ResNet101	trainval	33.1	50.3	–	11.6	34.9	51.2	–
CoupleNet [21]	ResNet101	trainval	34.4	54.8	37.2	13.4	38.1	50.8	–
Faster RCNN++ [42]	ResNet101-c4	trainval	34.9	55.7	37.4	15.6	38.7	50.9	3.36 s
Deformable RFCN [35]	ResNet101	trainval	34.5	55.0	–	14.0	37.7	50.3	125 ms
Attention CoupleNet [22]	ResNet101	trainval+S	35.4	55.7	37.6	13.2	38.6	52.5	–
Faster RCNN w/FPN [23]	ResNet101	trainval35k	36.2	59.1	39.0	18.2	39.0	48.2	240 ms
Single-stage									
SSD300† [5]	MobileNet	trainval35k	19.3	–	–	–	–	–	–
RFBNet300† [27]	MobileNet	trainval35k	20.7	–	–	–	–	–	–
YOLOv2 [26]	Darknet19	trainval35k	21.6	44.0	19.2	5.0	22.4	35.5	25 ms
SSD300 [5]	VGG16	trainval35k	25.1	43.1	25.8	6.6	25.9	41.4	12 ms
DSSD321 [25]	ResNet101	trainval35k	28.0	46.1	29.2	7.4	28.1	47.6	–
SSD512 [5]	VGG16	trainval35k	28.8	48.5	30.3	10.9	31.8	43.5	28 ms
RefineDet320 [6]	VGG16	trainval35k	29.4	49.2	31.3	10.0	32.0	44.4	–

(*Continued*)

TABLE 4.4 (Continued) Results on COCO test-dev

Method	Backbone	Train data	AP	AP@0.5	AP@0.75	AP_S	AP_M	AP_L	Time
RFBNet300 [27]	VGG16	trainval35k	30.3	49.3	31.8	11.8	31.9	45.9	15 ms
RefineDet320 [6]	ResNet101	trainval35k	32.0	51.4	34.2	10.5	34.7	50.4	–
SSD513 [25]	ResNet101	trainval35k	31.2	50.4	33.3	10.2	34.5	49.8	–
YOLOv3-608 [29]	DarkNet53	trainval35k	33.0	57.9	34.4	18.3	35.4	41.9	51 ms
RefineDet512 [6]	VGG16	trainval35k	**33.0**	**54.5**	**35.5**	**16.3**	**36.3**	**44.3**	–
DSSD513 [25]	ResNet101	trainval35k	33.2	53.3	35.2	13.0	35.4	51.1	182 ms
RetinaNet500 [24]	ResNet101	trainval35k	34.4	53.1	36.8	14.7	38.5	49.1	90 ms
RFBNet512 [27]	VGG16	trainval35k	**34.4**	**55.7**	**36.4**	**17.6**	**37.0**	**47.6**	33 ms
RefineDet512 [6]	ResNet101	trainval35k	36.4	57.5	39.5	16.6	39.9	51.4	–
GA-RetinaNet [37]	ResNet50	train	37.1	56.9	40.0	20.1	40.1	48.0	–
Cas-RetinaNet800 [36]	ResNet101	trainval35k	41.1	60.7	45.0	23.7	44.4	52.9	–
DRNet320⁺	MobileNet	trainval35k	26.0/25.7	45.3	26.8	8.0	28.7	38.9	27 ms
DRNet512⁺	MobileNet	trainval35k	28.5/28.4	49.8	29.6	14.3	32.1	36.6	51 ms
DRNet320	VGG16	trainval35k	30.5	51.2	32.3	11.2	33.9	44.9	29 ms
DRNet512	VGG16	trainval35k	**34.3**	**57.1**	**36.4**	**17.9**	**38.1**	**44.8**	53 ms
DRNet320	ResNet101	trainval35k	33.5	53.4	35.9	11.5	37.4	50.6	53 ms
DRNet320	ResNet101	trainval35k	38.6	60.3	42.2	19.0	43.2	52.7	36 ms
DRNet512	ResNet101	trainval35k	35.4	57.8	37.7	19.7	38.5	45.7	61 ms
DRNet320+	VGG16	trainval35k	37.9	61.6	40.3	22.6	40.4	48.4	–
DRNet512+	VGG16	trainval35k	39.2	61.3	42.3	21.4	42.8	51.4	–
DRNet320+	RedNet101	trainval35k	42.4	65.5	46.1	25.7	45.3	55.0	–

"AP" is evaluated at IoU thresholds from 0.5 to 0.95. "AP@0.5": PASCAL-type metric, IoU = 0.5. "AP@0.75": evaluated at IoU = 0.75. AP_S, AP_M, AP_L: AP at different scales. "+" indicates multi-scale testing. Bold fonts indicate that we draw readers' attention to a deep comparison AP at different scales. "+" indicates multi-scale testing. Bold fonts indicate that we draw readers' attention to a deep comparison

⁺: Prior MobileNet-based models are tested on COCO minival2014, so our MobileNet-based AP is reported as "test-dev/minival2014."

DRNet outperforms Faster RCNN, SSD, and RFBNet by a substantial margin. DRNet and RFBNet are similar in VOC mAP and COCO AP. Although similar performances are produced, DRNet and RFBNet are designed based on different motivations. That is, DRNet solves the problem of inaccurate anchors and feature locations, while RFBNet enhances the receptive field of the backbone. Two ideas are complementary so that anchor-offset detection and RFB block can be employed simultaneously.

Error analysis of DRNet512 is conducted on COCO 2014 minival set (5,000 images), and precision-recall curves are shown on person, vehicle, furniture, and electronic classes. From Figure 4.8, it is seen that there exists room for improvement of location precision. As for classification, DRNet has less confusion with similar categories or others (Sim & Oth). Thus, our approach is good at inter-class inference, benefiting from

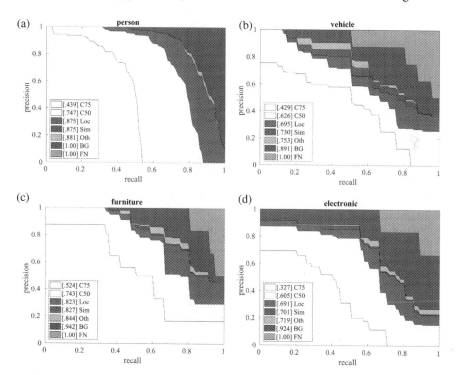

FIGURE 4.8 Error analysis of DRNet512 on person (a), vehicle (b), furniture (c), and electronic (d) classes in the COCO 2014 minival set. Each sub-figure shows the cumulative fraction of detections that are correct (Cor) or false positive due to poor localization (Loc), confusion with similar categories (Sim), with others (Oth), or with background (BG).

accurate single-stage detection features generated by the feature location refinement. By contrast, the error caused by the background (BG) is slightly serious. Probable improvement proposals will be discussed in Section 4.5.7.

4.5.5 Results on ImageNet VID

TRNet and TDRNet are evaluated on ImageNet VID dataset, which requires algorithms to detect 30-class targets in consecutive frames. The initial learning rate is 0.001 for the first 70 epochs, then we use a learning rate of 0.0001 for the next 30 epochs and 0.00001 for another 30 epochs. For fast inference speed, all models use 320×320 input images.

4.5.5.1 Accuracy vs. Speed Trade-off

SSD with four-scale detection features serves as the baseline, called SSD4s, and RG and RD are also contrasted with similar structure (see Figure 4.5a). As a result, SSD4s-VGG16 and SSD4s-MobileNet obtain 63.0% and 58.3% in mAP, respectively. The key frame duration is used for temporal detection, so accuracy vs. speed trade-off based on k is first analyzed. As shown in Figure 4.9a, TRNet significantly improves the mAP by 3.6% (i.e., 66.6% vs. 63.0). As k increases, the mAP decreases while the speed raises. Note that NMS impacts detection speed to some extent, but this part is out of the scope of this chapter, so the FPS without NMS is also reported (denoted as *Forward FPS*). As plotted in Figure 4.9a, *Forward FPS* increases from 136.8 to 234.2 with the rise of k, and the overall speed can reach 55.5 FPS. Furthermore, TDRNet improves the performance up to 67.5% benefiting from the proposed anchor-offset detection, which outperforms the baseline by 4.5 points. When $k = 8$, TDRNet can run at 55.1 FPS (*Forward FPS* reaches 215.4) while maintaining the mAP of 66.6%. As for $k = 1, 2, \ldots, 8$, TRNet has a 2.6-point drop in mAP (i.e., 66.6% vs. 64.0%), whereas TDRNet only has a decrease of 0.9% mAP (i.e., 67.5% vs. 66.6%). Thus, the refinement information in TDRNet is more robust in terms of temporal propagation owing to our proposed anchor-offset detection. Additionally, using MobileNet as the backbone, TRNet and TDRNet achieve 60.7% and 63.1% mAP ($k = 4$), surpassing the baseline by 2.4 and 4.8 points, respectively. Meanwhile, our MobileNet-based model can run over 70 FPS. (Note that the speed of MobileNet in Pytorch is slightly slower than the official implementation.)

To overcome TRNet's rapid mAP decrease with increasing k, the soft refinement strategy is introduced to TRNet with a soft coefficient e.

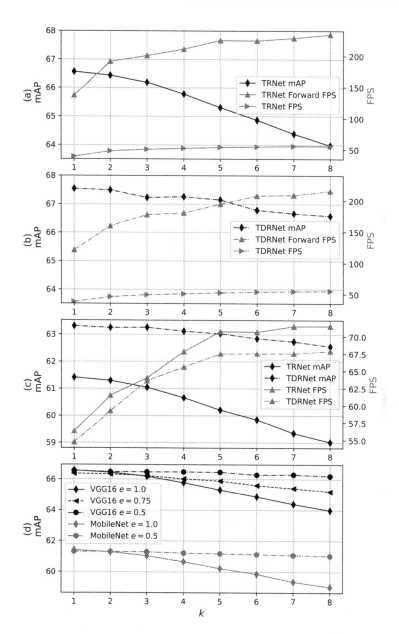

FIGURE 4.9 Inference analysis of TRNet and TDRNet with detection period k and soft coefficient e. *Forward FPS* does not consider the time consumption caused by NMS. (a) TRNet-VGG16 (baseline is 63.0%). (b) TDRNet-VGG16 (baseline is 63.0%). (c) TRNet and TDRNet with MobileNet (baseline is 58.3%). (d) TRNet with VGG16 and MobileNet.

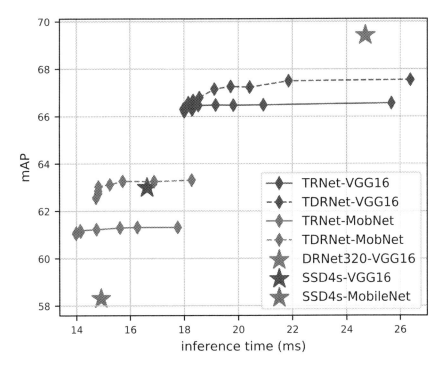

FIGURE 4.10 The plot of mAP vs. inference time for approaches in this chapter. We achieve a wide variety of trade-offs between accuracy and speed through different backbones and k settings.

Referring to Figure 4.9d, $e=0.5$ can restrict this mAP drop within 1% from $k=1$ to $k=8$, i.e., 66.6% vs. 66.2% for VGG16 and 61.3% vs. 61.0% for MobileNet.

As shown in Figure 4.10, with 320×320 input size, this chapter presents a series of approaches for the trade-off between accuracy and speed. The fast solution is TRNet-MobileNet ($k=8$), whose inference time is 14 ms. The most accurate method in this chapter is DRNet320, with 69.4% mAP and 25 ms in inference. In terms of TDRNet-VGG16 and DRNet320-VGG16, it can be seen that DRNet is more accurate owing to FPN and multi-deformable head, yet TDRNet has a better trade-off between accuracy and speed.

4.5.5.2 Comparison with Other Architectures
TRNet and TDRNet are compared against several prior and contemporary approaches in Table 4.5. Existing video detectors are categorized

TABLE 4.5   Comparison of the proposed methods and several prior and contemporary approaches on VID

| Method | Backbone | Components | | | Performances | |
		Flow	Tracking	RNN	Real time	mAP
Static methods						
SSD4s320	MobileNet				√	58.3
SSD4s320	VGG16				√	63.0
Faster RCNN [2]	GoogleNet					63.0
SSD [6]	VGG16				√	63.0
RefineDet320 [6]	VGG16				√	66.7
DRNet320	VGG16				√	69.4
Offline methods						
STMN [14]	VGG16			√		55.6
TPN [17]	GoogLeNet			√		68.4
FGFA [10]	ResNet101	√				76.3
HPVD [11]	ResNet101	√				78.6
STSN [13]	ResNet101					78.7
D&T [9]	ResNet101		√			79.8
Online methods						
LSTM-SSD [15]	MobileNet			√	√	54.5
HPVD-Mob [12]	MobileNet	√		√	√	60.2
TCNN [8]	DeepID+Craft	√	√		√	61.5
TSSD [16]	VGG16			√	√	65.4
TRNet	MobileNet				√	61.2
TDRNet	MobileNet				√	63.1
TRNet	VGG16				√	66.5
TDRNet	VGG16				√	67.3

$k = 4$ for TRNet and TDRNet.

into offline methods (i.e., batch-frame mode) and online methods. Most methods are based on a two-stage detector and a deep backbone, so they usually have a high mAP yet impractical execution time. As for offline approaches, this non-causal batch-frame mode usually leverages both previous and future information that prohibits it from real-world applications. In addition, recent works usually borrow other temporal modules (e.g., tracker, optical flow, and LSTM) to integrate multi-frame information. Among single-stage methods, TDRNet-VGG16 has a significant superiority in accuracy, i.e., 1.9% and 12.9% higher mAP than TSSD [17] and LSTM-SSD [16], respectively. When compared to MobileNet-based detectors, TDRNet-MobileNet has the best results, i.e., it outperforms LSTM-SSD by 8.7 points and surpasses HPVD-Mob [13] by 2.9 points. To the best of our knowledge, our designs have the following merits: (1) Instead of borrowing other temporal modules, temporal information is exploited from the detector itself. Thus, our design is a new online detection mode for videos; (2) TDRNet achieves the highest mAP among real-time online temporal detectors, and it induces a better trade-off between accuracy and speed for real-world tasks.

4.5.6 Discussion

4.5.6.1 Key Frame Scheduling

Zhu et al. developed an adaptive key frame scheduling [12] for key frame selection, and we employ a prefixed key frame duration. We argue that the adaptive key frame scheduling is needless for both accuracy and speed in this chapter: (1) Any scheduling strategy cannot generate an mAP that outmatches the result of $k = 1$. Therefore, given that the speed of RG is fast enough (i.e., 270 FPS) and a scheduling strategy should deal with each frame, it is better to set $k = 1$ than to use an adaptive key frame scheduling. (2) The longest period is $k = 8$ in our experiments, and we also state that longer detection period is needless: *Forward FPS* of SSD4s-VGG16 is 270 and that of TRNet-VGG16 reaches 234 ($k = 8$). Thus, longer key frame duration has an ignorable contribution to inference speed since TRNet and TDRNet cannot overpass SSD4s in *Forward FPS*.

4.5.6.2 Further Enhancement of Refinement Networks

In terms of accuracy, it can be seen from Figure 4.8 that there still exists room for improvement of location precision and foreground-background classification. We present two probable solutions: (1) Multistep refinement could be beneficial; (2) because of the hard negative mining, only a part of negative samples (i.e., background) are used for training. Therefore, using a focal loss [23] to train a network with all negative samples could be more effective. For example, Chi et al. used focal loss and negative anchor filtering to train a refinement network and achieved high performance on face detection [43].

Regarding inference speed, the NMS has an impact. There could exist two solutions: (1) Decreasing the anchor amount could be beneficial; (2) an end-to-end detector is becoming urgently necessary. For example,

Hu et al. developed a relation network for both detection and duplicate removal so that the whole network can perform in an end-to-end manner [44].

4.5.6.3 Refinement Networks for Real-World Object Detection

Object detection is the foundation of robotic scene perception. Different from datasets, real-world detection has two main challenges: Few-shot condition and domain shift problem. As for few-shot detection, the refinement network and anchor-offset detection can be combined with metric learning [45] to solve few-shot problem. In terms of domain shift, underwater scenes have a wide variety of data domain, which can be solved by the combination of visual restoration (see Chapter 2 of this book) and detection. The analysis of the domain shift problem will be presented in Chapter 6 of this book.

4.6 CONCLUDING REMARKS

In this chapter, we have taken aim at precisely detecting objects in real time for static and temporal scenes. Firstly, drawbacks of the single-stage detector are analyzed from the strengths of two-stage methods. Thereby, including an anchor refinement, a feature location refinement, and a deformable detection head, a novel anchor-offset detection is proposed. Besides two-step regression, the anchor-offset detection is also able to capture accurate single-stage features for detection. Correspondingly, a DRNet is proposed based on the anchor-offset detection, where a multi-deformable head is also designed for more contextual information. In the case of temporal detection, we propagate the refinement information in the anchor-offset detection across time and propose a TRNet and a TDRNet with a reference generator and a refinement detector. Our developed approaches have been evaluated on PASCAL VOC, COCO, and ImageNet VID. As a result, our designs induce a considerably enhanced detection accuracy and see a substantial improvement on the trade-off between accuracy and speed. Finally, the proposed algorithms are applied to underwater object detection and grasping.

In the future, we plan to incorporate attention mechanism to the anchor-offset detection and design more effective networks for more robust feature learning. In addition, the idea of refinement will be used for few-shot detection and real-world detection problem.

REFERENCES

1. Z. Zhao, P. Zheng, S. Xu, and X. Wu, "Object detection with deep learning: A review," *IEEE Trans. Neural Networks Learn. Syst.*, vol. 30, no. 11, pp. 3212–3232, 2019.

2. S. Ren, K. He, R. Girshick, and J. Sun, "Faster R-CNN: Towards realtime object detection with region proposal networks," in *Proceedings of the Advances in Neural Information Processing Systems*, Montreal, Canada, Dec. 2015, pp. 91–99.

3. J. Dai, Y. Li, K. He, and J. Sun, "R-FCN: Object detection via regionbased fully convolutional networks," in *Proceedings of the Advances in Neural Information Processing Systems*, Barcelona, Spain, Dec. 2016, pp. 379–387.

4. J. Redmon, S. Divvala, R. Girshick, and A. Farhadi, "You only look once: Unified, real-time object detection," in *Proceedings of the IEEE Conference on Computer Vision and Pattern Recognition*, Las Vegas, USA, Jun. 2016, pp. 779–788.

5. W. Liu, D. Anguelov, D. Erhan, C. Szegedy, S. Reed, C. Y. Fu, and A. C. Berg, "SSD: Single shot multibox detector," in *Proceedings of the European Conference on Computer Vision*, Amsterdam, Netherlands, Oct. 2016, pp. 21–37.

6. S. Zhang, L. Wen, X. Bian, Z. Lei, and S. Z. Li, "Single-shot refinement neural network for object detection," in *Proceedings of the IEEE Conference on Computer. Vision and Pattern Recognition*, Salt Lake City, USA, Jun. 2018, pp. 4203–4212.

7. O. Russakovsky, J. Deng, H. Su, J. Krause, S. Satheesh, S. Ma, Z. Huang, A. Karpathy, A. Khosla, M. Bernstein, A. C. Berg, and F. Li, "ImageNet large scale visual recognition challenge," *Int. J. Comput. Vision*, vol. 115, no. 3, pp. 211–252, 2015.

8. W. Han, P. Khorrami, T. L. Paine, P. Ramachandran, M. Babaeizadeh, H. Shi, J. Li, S. Yan, and T. S. Huang, "Seq-NMS for video object detection," https://arxiv.org/abs/1602.08465, 2016.

9. K. Kang, H. Li, J. Yan, X. Zeng, B. Yang, T. Xiao, C. Zhang, Z. Wang, R. Wang, X. Wang, and X. Ouyang, "T-CNN: Tubelets with convolutional neural networks for object detection from videos," *IEEE Trans. Circuits Syst. Video Technol.*, vol. 28, no. 10, pp. 2896–2907, 2018.

10. C. Feichtenhofer, A. Pinz, and A. Zisserman, "Detect to track and track to detect," in *Proceedings of the International Conference on Computer Vision*, Venice, Italy, Oct. 2017, pp. 3038–3046.

11. X. Zhu, Y. Wang, J. Dai, L. Yuan, and Y. Wei, "Flow-guided feature aggregation for video object detection," in *Proceedings of the International Conference on Computer Vision*, Venice, Italy, Oct. 2017, pp. 408–417.

12. X. Zhu, J. Dai, L. Yuan, and Y. Wei, "Towards high performance video object detection," in *Proceedings of the IEEE Conference on Computer Vision and Pattern Recognition*, Salt Lake City, USA, Jun. 2018, pp. 7210–7218.

13. X. Zhu, J. Dai, X. Zhu, Y. Wei, and L. Yuan, "Towards high performance video object detection for mobiles," https://arxiv.org/abs/1804.05830, 2018.

14. G. Bertasius, L. Torresani, and J. Shi, "Object detection in video with spatiotemporal sampling networks," in *Proceedings of the European Conference on Computer Vision*, Munich, Germany, Sep. 2018, pp. 342–357.

15. F. Xiao and Y. J. Lee, "Video object detection with an aligned spatialtemporal memory," in *Proceedings of the European Conference on Computer Vision*, Munich, Germany, Sep. 2018, pp. 494–510.

16. M. Liu and M. Zhu, "Mobile video object detection with temporally aware feature maps," in *Proceedings of the IEEE Conference on Computer Vision and Pattern Recognition*, Salt Lake City, USA, Jun. 2018, pp. 5686–5695.

17. X. Chen, J. Yu, and Z. Wu, "Temporally identity-aware SSD with attentional LSTM," *IEEE Trans. Cybern.*, vol. 50, no. 6, pp. 2674–2686, 2020.

18. K. Kang, H. Li, T. Xiao, W. Ouyang, J. Yan, X. Liu, and X. Wang, "Object detection in videos with tubelet proposal networks," in *Proceedings of the IEEE Conference on Computer Vision and Pattern Recognition*, Honolulu, USA, Jul. 2017, pp. 727–735.

19. M. Everingham, L. Van Gool, C. K. Williams, J. Winn, and A. Zisserman, "The pascal visual object classes (VOC) challenge," *Int. J. Comput. Vision*, vol. 88, no. 2, pp. 303–338, 2010.

20. T. Y. Lin, M. Maire, S. Belongie, J. Hays, P. Perona, D. Ramanan, P. Dollar, and C. L. Zitnick, "Microsoft coco: Common objects in´ context," in *Proceedings of the European Conference on Computer Vision*, Zurich, Switzerland, Sep. 2014, pp. 740–755, 2014.

21. Y. Zhu, C. Zhao, J. Wang, X. Zhao, Y. Wu, H. Lu, "Couplenet: Coupling global structure with local parts for object detection," in *Proceedings of the IEEE International Conference on Computer Vision*, Venice, Italy, Aug. 2017, pp. 4126–4134.

22. Y. Zhu, C. Zhao, H. Guo, J. Wang, X. Zhao, and H. Lu, "Attention couplenet: Fully convolutional attention coupling network for object detection," *IEEE Trans. Image Process.*, vol. 28, no. 1, pp. 113–126, 2019.

23. T. Y. Lin, P. Goyal, R. Girshick, K. He, and P. Dollar, "Focal loss for dense object detection," in *Proceedings of the IEEE International Conference on Computer Vision*, Venice, Italy, Oct. 2017, pp. 2980–2988.

24. C. Fu, W. Liu, A. Ranga, A. Tyagi, and A. C. Berg, "DSSD: Deconvolutional single shot detector," https://arxiv.org/abs/1701.06659, 2017.

25. J. Redmon and A. Farhadi, "YOLO9000: Better, faster, stronger," in *Proceedings of the IEEE Conference on Computer Vision and Pattern Recognition*, Honolulu, USA, Jul., 2017, pp. 6517–6525.

26. S. Liu, D. Huang, and Y. Wang, "Receptive field block net for accurate and fast object detection," in *Proceedings of the European Conference on Computer Vision*, Munich, Germany, Sep. 2018, pp. 404–419.

27. Z. Shen, Z. Liu, J. Li, Y. Jiang, Y. Chen, and X. Xue, "DSOD: Learning deeply supervised object detectors from scratch," in *Proceedings of the IEEE International Conference on Computer Vision*, Venice, Italy, Oct. 2017, pp. 1919–1927.

28. J. Redmon and A. Farhadi, "YOLOv3: An incremental improvement," https://arxiv.org/abs/1804.02767, 2018.

29. T. Y. Lin, P. Dollar, R. Girshick, K. He, B. Hariharan, S. Belongie, "Feature pyramid networks for object detection," in *Proceedings of the IEEE Conference on Computer Vision and Pattern Recognition*, Honolulu, USA, Jun. 2017, pp. 2117–2125.

30. Y. Pang, J. Cao, and X. Li, "Learning sampling distributions for efficient object detection," *IEEE Trans. Cybern.*, vol. 47, no. 1, pp. 117–129, 2017.

31. J. Dai H. Qi, Y. Xiong, Y. Li, G. Zhang, H. Hu, and Y. Wei, "Deformable convolutional networks," in *Proceedings of the International Conference on Computer Vision*, Venice, Italy, Oct. 2017, pp.764–773.

32. H. Zhang, H. Chang, B. Ma, S. Shan, X. Chen, "Cascade retinanet: Maintaining consistency for single-stage object detection," in *Proceedings of the British Machine Vision Conference*, Cardiff, UK, Sep. 2019, pp. 1–12.

33. J. Wang, K. Chen, S. Yang, C. C. Loy, and D. Lin, "Region proposal by guided anchoring," in *Proceedings of the IEEE Conference on Computer Vision and Pattern Recognition*, Long Beach, USA, 2019, pp. 2965–2974.

34. K. Simonyan and A. Zisserman, "Very deep convolutional networks for large-scale image recognition," https://arxiv.org/abs/1409.1556, 2014.

35. S. Ioffe, and C. Szegedy, "Batch normalization: Accelerating deep network training by reducing internal covariate shift," in *Proceedings of the International Conference on Machine Learning*, Lile, France, Jul. 2015, pp. 448–456.

36. K. He, X. Zhang, S. Ren, and J. Sun, "Deep residual learning for image recognition," in *Proceedings of the IEEE Conference on Computer Vision and Pattern Recognition*, Las Vegas, USA, Jun. 2016, pp. 770–778.

37. H. Law and J. Deng, "CornerNet: Detecting objects as paired keypoints," in *Proceedings of the European Conference on Computer Vision*, Munich, Germany, Sep. 2018, pp. 734–750.

38. X. Zhou, J. Zhuo, and P. Krahenb, "Bottom-up object detection by grouping extreme and center points," in *Proceedings of the IEEE Conference on Computer Vision and Pattern Recognition*, Long Beach, USA, Jun. 2019, pp. 850–859.

39. X. Zhou, D. Wang, and P. Krahenb, "Objects as points," https://arxiv.org/abs/1904.07850, 2019.

40. Z. Tian, C. Shen, H. Chen, T. He, "FCOS: Fully convolutional one-stage object detection," in *Proceedings of the IEEE/CVF International Conference on Computer Vision (ICCV)*, Seoul, Korea, Oct. 2019, pp. 9627–9636. *arXiv:1904.01355*, 2019.

41. W. Luo, Y. Li, R. Urtasun, and R. Zemel, "Understanding the effective receptive field in deep convolutional neural networks," in *Proceedings of the Advances in Neural Information Processing Systems*, Barcelona, Spain, Dec. 2016, pp. 4898–4906.

42. A. Howard, M. Zhu, B. Chen, D. Kalenichenko, W. Wang, T. Weyand, M. Andreetto, and H. Adam, "Mobilenets: Efficient convolutional neural

networks for mobile vision applications," https://arxiv.org/abs/1704.04861, 2017.

43. C. Chi, S. Zhang, J. Xing, Z. Lei, S. Z. Li, and X. Zou, "Selective refinement network for high performance face detection," in *Proceedings of the AAAI Conference on Artificial Intelligence*, Honolulu, USA, Jan. 2019, pp. 8231–8238.

44. H. Hu, J. Gu, Z. Zhang, J. Dai, Y. Wei, "Relation networks for object detection," in *Proceedings of the IEEE Conference on Computer Vision and Pattern Recognition*, Salt Lake City, USA, Jun. 2018, pp. 3588–3597.

45. L. Karlinsky, J. Shtok, S. Harary, E. Schwartz, A. Aides, R. Feris, R. Giryes, "A. M. Bronstein: RepMet: Representative-based metric learning for classification and few-shot object detection," in *Proceedings of the IEEE Conference on Computer Vision and Pattern Recognition*, Long Beach, USA, Jun. 2019, 5197–5206.

CHAPTER 5

Rethinking Temporal Object Detection from Robotic Perspectives

5.1 INTRODUCTION

As a temporal perception task, video object detection (VID) is one of the main research areas in computer vision and a fundamental tactic for robotic perception. Over recent years, we have witnessed the development of temporal object detection on ImageNet VID dataset [1]. As the exclusive assessment, average precision (AP) has grown from less than 50% [2] to over 80% [3].

Different from images, video frames are interrelated on time series. Thus, almost all VID methods leverage across-time information to improve detection performance [3–11], but their metric AP only statically reflects image-based accuracy, failing in across-time evaluation. We deem that the AP would neglect some important characteristics in VID. As shown in Figure 5.1a,b, besides false positive/negative captured by AP, defective cases of temporal detection are subdivided into twofold aspects:

- *Recall continuity*: As shown in Figure 5.1a, transient object recall induces short tracklet duration, whose duration could only contain several frames. Additionally, intermittent object missing forms

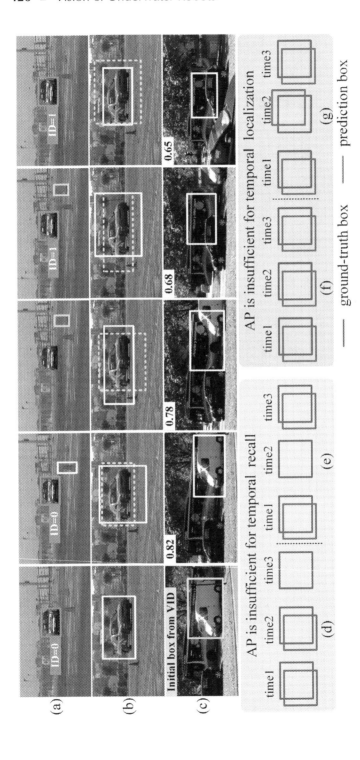

FIGURE 5.1 Defective detection and tracking cases. (a) Short tracklet duration (white tracklet only contains one object while the length of light gray one is two) and tracklet fragment; (b) box center/size jitter (light gray dashed boxes denote previous results); (c) Siamese tracker needs an initial box from VID and is prone to drift (tracking score is shown in the left-top); (d–g) AP can hardly describe temporal recall/localization.

tracklet fragments, which could incur identity switch. We deem these phenomena damage recall continuity in VID.

- *Localization stability*: As shown in Figure 5.1b, box center/size jitter frequently appears in modern object detection, and slight pixel-level change could incur considerable location jitter. It is conceived that this phenomenon impairs localization stability in VID.

Compared to detection accuracy, continuity and stability are equally important in robotic perception. Nevertheless, there has been relatively little work studying these two problems because the static evaluation system has difficulty in reflecting them. That is, AP performs not so well on describing detection continuity and stability. On the one hand, although it reports object recall/missing from a spatial perspective, AP is insufficient for analysis of temporal classification. For example, AP can hardly distinguish cases in Figure 5.1d,e, but Figure 5.1d is relatively better as for robotic perception. On the other hand, AP is also insufficient for localization jitter since intersection-over-union (IoU) cannot describe the direction of overlap. That is, AP is unable to discriminate cases in Figure 5.1f,g, but Figure 5.1f is relatively better as for temporal localization.

Above problems inspire us to develop a temporal evaluation system for VID. We suggest a non-reference manner for evaluation because (1) both continuity and stability are totally unrelated to man-made labels and (2) the evaluation system should be applied to various scenes (including but not limited to datasets), and most robotic scenarios lack ground truth labels. As a relevant scope, multi-object tracking (MOT) aims to associate object boxes across time [12] and then object tracklets are produced. The tracklets from MOT exactly reflect VID performance across time, but there has been limited research on MOT-based VID analysis. This motivates us to reinvestigate VID approaches and then generally evaluate and enhance detectors' continuity/stability with MOT.

VID-MOT deals with multiple tracklets in the manner of tracking-by-detection, but robotic perception also has an imperative need for single object tracking (SOT) [13]. If VID-MOT is directly leveraged for SOT, redundant computing would incur high time consumption. Additionally, researchers tend to exploit similarity-based SOT methods [14–17].

However, as shown in Figure 5.1c, similarity-based SOT is independent of VID and has threefold drawbacks: (1) It is prone to suffer tracking drift; (2) it cannot work without initial box from VID [17]; (3) VID-SOT cascade would highly increase model complexity [18]. All these limitations are unfavorable for robotic perception, inspiring us to extend VID methods toward SOT task.

In this chapter, we firstly study temporal performance on object detection. We pose transient and intermittent object recall/missing as a problem of recall continuity, while box center/size jitter is formulated as an issue of localization stability. Further, performing favorably in robotic scenarios, novel non-reference assessments are proposed based on MOT rather than ground truth labels. To this end, we modify MOT pipeline to capture recall failure (i.e., object missing) and design a Fourier approach for stability evaluation. In addition, including short tracklet suppression, fragment filling, and temporal location fusion, online tracklet refinement (OTR) is proposed to enhance VID continuity and stability. The proposed OTR can be generally applied to any detector in temporal tasks. Subsequently, we design small-overlap suppression (SOS) for extending VID approaches to SOT task, and thus SOT-by-detection is proposed. Compared to VID task, the SOS is able to induce a faster inference speed for SOT task. Compared to similarity-based SOT, the proposed SOT-by-detection has advantages on free initialization and flexibility. Our contributions are summarized as follows:

- Two VID problems are novelly analyzed from the robotic perspective, i.e., continuity and stability and then non-reference assessments are proposed for them. Our assessments can make up the deficiency of traditional accuracy-based evaluation.

- We propose an OTR to generally improve detection continuity/stability. We also discuss how to solve these problems with the detector itself to inspire future works.

- We propose an SOS to extend VID approaches to the SOT task without the requirement of a similarity-based SOT module. The proposed SOT-by-detection is flexible for VID, MOT, and SOT tasks in robotic perception.

The remainder of this chapter is organized as follows. Section 5.2 presents related works. Non-reference assessments for temporal performance evaluation and OTR are elaborated in Section 5.3. Section 5.4 presents SOT-by-detection in detail, and Section 5.5 provides the experimental results and discussion. Conclusions are summarized in Section 5.6.

5.2 REVIEW OF TEMPORAL DETECTION AND TRACKING

5.2.1 Temporal Object Detection

There are manifold ideas of temporal detection, including (1) post-processing [4], (2) tracking-based location [3, 5, 6, 18], (3) feature aggregation [7–9], (4) batch-frame processing [10], and (5) temporally sustained proposal [11]. All these ideas are attractive in that they can leverage temporal information for detection.

All the above methods pursued high accuracy and followed AP evaluation. However, detection continuity and stability are equally important in robotic cases. AP considers static accuracy based on detection recall rate and precision [19], but it can hardly give a temporal evaluation for VID methods as mentioned in Section 5.1. Zhang and Wang proposed evaluation metrics for VID stability and proved that the stability has a low correlation with AP [20]. In detail, they formulated the stability problem as fragment error, center position error, and scale/ratio error. Their work was impressive but had two limitations: (1) They ignored the problem of short tracklet duration that was also an import situation in recall continuity; (2) their evaluation needed ground truth boxes, hampering their metrics from extensive applications. Conversely, we address these limitations and propose non-reference assessments without the need of man-made labels.

5.2.2 Tracking Metrics

Referring to [21], MOT metrics included multi-object tracking accuracy (MOTA) and precision (MOTP). MOTA synthesized false positive, false negative, and identity switch of detected objects while MOTP considered the static localization precision. Therefore, tracklet fragment was captured by identity switch, but some other tracklet characteristics were ignored (e.g., short tracklet duration).

VOT used expected average overlap rate (EAO) for SOT evaluation [13], including accuracy and robustness. Accuracy was determined by static IoU, and the robustness described tracking failure. After tracking failure was captured by the evaluation process, the tracker would be initialized with ground truth. EAO could describe tracking fragments, but it could not give a comprehensive evaluation for multi-tracklets.

Tracking metrics (i.e., MOTA, MOTP, and EAO) are able to describe temporal recall, but similar to AP, they are insufficient for evaluating temporal localization. Moreover, all existing evaluations are based on ground truth. In contrast, we propose a Fourier approach to directly describe box jitter without the need for labels.

5.2.3 Tracking-by-Detection (i.e., MOT)

MOT methods associate detected boxes across time. For example, Bewley et al. leveraged Kalman and Hungarian methods for fast MOT [22]. Lu et al. exploited RNN for sequential modeling and contrasted MOT approach with LSTM [23]. Almost all these methods are based on VID, but the effect of MOT on VID is usually ignored. Analytically, we report the MOT-based VID evaluation and enhancement. In addition, we exploit SOT-by-detection for flexible object perception.

5.2.4 Detection-SOT Cascade

Detection and SOT are distinct in their pipelines, and researchers tried to simultaneously leverage their advantages. Kang et al. used a tracking algorithm to rescore detection results with around tracklets [6]. Kim et al. combined a detector, a forward tracker, and a backward tracker to perform tracking-detection refinement [18]. Feichtenhofer et al. simultaneously leveraged a two-stage detector and a correlation filter to boost VID accuracy [3]. Luo et al. formulated detection and tracking as a sequential decision problem and processed a frame by either a siamese tracker or a detector [5]. These methods complementally improved tracking and detection, but their SOT model and detector are independent so that high model complexity is usually incurred. Instead of the model cascade, we design an SOS to extend detection methods toward SOT task, producing SOT-by-detection framework.

5.3 ON VID TEMPORAL PERFORMANCE

5.3.1 Non-reference Assessments

Our assessments follow a reasonable assumption, i.e., object motion is smooth across time without high-frequency location jitter or change of existence. We leverage a detector and an MOT module to recall all object tracklets in a video. In detail, any detector to be evaluated can be used for VID, and we employ the IoU-based MOT tracker reported by [9] to associate detected boxes. Unlike label-based evaluation, we only focus on detected tracklets, because totally missed tracklet does not impact continuity and stability.

As delineated in Figure 5.2, a detector locates and classifies objects at each frame f. Each object has confidence score s, box center c_x, c_y, and size w, h. N tracklets $\{T_n \,|\, n = 1, 2, \ldots, N\}$ are produced after the whole video is processed by VID and MOT. The video duration and the tracklet duration are denoted as t_v, t_n. The vertical axis of Figure 5.2 describes c_x, c_{y}, w, or h.

5.3.1.1 Recall Continuity

Object tracklets should have an appropriate duration without interruption, and transient/intermittent object recall/missing damages recall continuity. As for this problem, we consider the impact of short tracklet duration and tracklet fragment. Referring to Figure 5.1a and T_2 in Figure 5.2, tracklets with short duration frequently appear in VID. To capture them, we design extremely short duration error (ESDE) and short duration error (SDE) with various duration thresholds as follows:

$$\text{ESDE/SDE} = \frac{1}{t_v} \sum_{n=1}^{N} (t_n \quad \text{if } t_n < S \text{ else } 0). \tag{5.1}$$

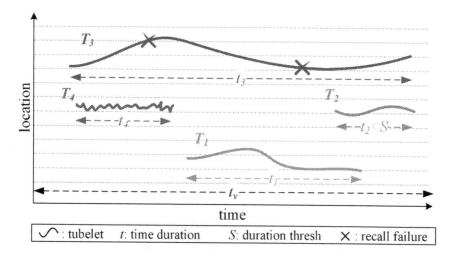

FIGURE 5.2 Problem formation. Object tracklets could suffer from short tracklet duration (e.g., T2), fragments (e.g., T3), location jitter (e.g., T4).

In this chapter, $S_{ESDE} = 3$, $S_{SDE} = 10$, which describe different degrees of short duration problem.

Figure 5.1a and $\mathcal{T}_3$ in Figure 5.2 describe tracklet fragment problem. Some MOT algorithms end a tracklet after a recall failure. Conversely, we count the number of continuous recall failure with S_{lost}, and leave a S_{lost}^{max} -frame life duration for each tracklet. That is, if a recall-failed tracklet is re-matched by a box in consequent S_{lost}^{max} frames, the tracklet can be kept. In this way, the total number of object missing om in the whole tracklet can be captured, forming tracklet fragment error (TFE) and fragmental tracklet ratio (FTR) as follows:

$$
\mathrm{TFE} = \sum_{n=1}^{N} om_n \bigg/ \sum_{n=1}^{N} t_n
$$

$$
\mathrm{FTR} = \frac{1}{N} \sum_{n=1}^{N} (1 \quad \text{if} \quad om_n > 0 \quad \text{else} \quad 0).
$$

(5.2)

TFE describes the ratio of object missing in tracklets, while FTR gives the ratio of faulty tracklets. They are complementary for tracklet fragment problem, i.e., a better VID result should have lower TFE and FTR in the meantime. That is, there is a small number of objects missing, and object missing is concentrated in a small number of tracklets. Note that these calculations are numerically small, so a log transformation is used to enhance the contrast, i.e., $\log_{100}(1 + 99 \times \alpha)$, where α represents ESDE, SDE, TFE, or FTR. Finally, recall continuity error (RCE) is defined as RCE = ESDE + SDE + TFE + FTR.

5.3.1.2 Localization Stability

Object tracklets should be smooth in localization, and box center/size jitter damages localization stability (see Figure 5.1b and $\mathcal{T}_4$ in Figure 5.2). We evaluate temporal stability in Fourier domain so that our approach can work without labels. Time domain data p can be transformed to Fourier domain by $P = \mathcal{F}(p)$, where p represents c_x, c_y, w, or h. Thus, P contains frequency information of p, and we extract frequency-related amplitude with $\tilde{P} = \mathrm{Abs}(P)$. Note that each tracklet produces different frequency component because of variable data length (i.e., tracklet duration). That is, $\tilde{P} = \{(q_k^p, A_k^p)\}, q_k^p = k/t, k = \{0, 1, \ldots, \lfloor t/2 \rfloor\}$. Here, q is the frequency set; t is tracklet duration; and A denotes frequency-related amplitude. Based on Fourier analysis, center jitter error (CJE) and size jitter error (SJE) are designed as

$$\mathrm{CJE} = \left(10^3 \sum_{n=1}^{N} \sum_{p \in \{c_x, c_y\}} \sum_{k=1}^{\lfloor t_n/2 \rfloor} q_{n,k}^p A_{n,k}^p \right) \bigg/ \sum_{n=1}^{N} t_n$$

$$\mathrm{SJE} = \left(10^3 \sum_{n=1}^{N} \sum_{p \in \{w, h\}} \sum_{k=1}^{\lfloor t_n/2 \rfloor} q_{n,k}^p A_{n,k}^p \right) \bigg/ \sum_{n=1}^{N} t_n$$

(5.3)

Ultimately, localization jitter error $\mathrm{LJE} = \mathrm{CJE} + \mathrm{SJE}$.

5.3.2 Online Tracklet Refinement

For enhancing recall continuity and localization stability, we refine VID results based on tracklets. A new attribute is used to describe tracklets, i.e., current duration S_{dur}. Therefore, a tracklet can be formulated as $\mathcal{T} = (\mathcal{D}, ID, S_{lost}, S_{dur})$, where S_{dur} records tracklet duration at each time-stamp; S_{lost} has been explained in Section 5.3.1; ID denotes tracklet identity; $\mathcal{D}$ is the object set in the tracklet (i.e., $\{(s, c_x, c_y, w, h)\}$), and the length of $\mathcal{D}$ (i.e., S_{obj}) cannot exceed $S_{obj}^{max} = 5$. That is, if $S_{dur} > S_{obj}^{max}$, only the latest S_{obj}^{max} objects are preserved in $\mathcal{D}$.

5.3.2.1 Short Tracklet Suppression

For suppressing short tracklets and enhancing ESDE/SDE, we define a tracklet as reliable tracklet if $S_{dur} > S_{SDE}$ and then boxes in unreliable tracklets are suppressed. This manner is beneficial to continuity, and it has twofold effects on accuracy. Firstly, false positives could be suppressed, because their recall is usually inconsecutive across time so that a reliable tracklet is hard to form. Secondly, false negatives could be produced since an object would not be reported until it forms an S_{SDE}-length tracklet.

5.3.2.2 Fragment Filling

In terms of the fragment issue and TFE/FTR, we make up for the object missing in a tracklet based on a reasonable assumption, i.e., the object motion is uniform in an extremely short duration (e.g., S_{obj}^{max}). When a tracklet suffers from a recall failure at the fth frame, its previous boxes $\{(c_x^{f-i}, c_y^{f-i}, w^{f-i}, h^{f-i}) \mid i = 1, 2, \ldots, S_{obj}\}$ can be used to predict current location. In detail, we first estimate the velocity v_p, i.e., $v_p = \left(\sum_{i=1}^{S_{obj}-1} p^{f-i} - p^{f-i-1} \right) \big/ (S_{obj} - 1)$ and then the current location can be given as $p = p^{f-1} + v_p$, where p denotes c_x, c_y, w, or h.

5.3.2.3 Temporal Location Fusion

For location stability and CJE/SJE, we add the object into its tracklet and then produce a new location with a weighted average. A geometric progression is contrasted with $\Omega = \{\omega^l \mid l = 1, \ldots, 0.1\}$, where l is an S_{obj}-length arithmetic progression. The normalized Ω is utilized as the weight to merge $\{p \mid p \in \mathcal{D} \,\& \, \mathcal{D} \in \mathcal{T}\}$, and the updated location can be formulated as $\check{p}^f = \sum_{i=0}^{S_{obj}} \Omega_i p^{f-i}$.

5.4 SOT-BY-DETECTION

5.4.1 Small-Overlap Suppression

Algorithm 1: SOS-NMS

Input: After selection by confidence threshold, boxes $\mathcal{B} = \{b_1, \ldots, b_m\}$, confidence scores $\mathcal{S} = \{s_1, \ldots, s_m\}$; previous tracked box b^{f-1}; SOS/NMS thresholds U^{sos}, U^{nms}.

Output: Tracked box b^f

begin:
　//SOS based on IoU ("+=/-=" denotes element add/removal)
　$\mathcal{B}^{sos} = \mathcal{B}; \mathcal{S}^{sos} = \mathcal{S}; \mathcal{O}^{sos} = iou(b^{f-1}, \mathcal{B})$
　for $(b_i, s_i, o_i) \in (\mathcal{B}^{sos}, \mathcal{S}^{sos}, \mathcal{I}o U^{sos})$ **do**
　　if $o_i < U^{sos}$ **then**
　　　$\mathcal{B}^{sos} -= b_i; \mathcal{S}^{sos} -= s_i; \mathcal{O}^{sos} -= o_i$
　　//Inspection of tracking failure
　　if $\mathcal{B}^{sos} = empty$ **then**
　　　return $b^f = empty$
　//NMS based on confidence score
　$\mathcal{B}^{nms} = \{\}; \mathcal{S}^{nms} = \{\}; \mathcal{O}^{nms} = \{\}$
　while $\mathcal{B}^{sos} \neq empty$ **do**
　　$idx = \mathrm{argmax}\mathcal{S}^{sos}$
　　$b = \mathcal{B}^{sos}_{idx}; s = \mathcal{S}^{sos}_{idx}; o = \mathcal{O}^{sos}_{idx}$
　　$\mathcal{B}^{nms} += b; \mathcal{S}^{nms} += s; \mathcal{O}^{nms} += o; \mathcal{B}^{sos} -= b; \mathcal{S}^{sos} -= s; \mathcal{O}^{sos} -= o$
　　if $iou(b, b_i) > U^{nms}$ **then**
　　　$\mathcal{B}^{sos} -= b_i; \mathcal{S}^{sos} -= s_i; \mathcal{O}^{sos} -= o_i$
　　//Selection of single box with IoU
　　$idx = \mathrm{argmax}\mathcal{O}^{nms}$
　　return $b^f = \mathcal{B}^{nms}_{idx}$

We promote VID model to generate SOT result by propagating the previous location $b^{f-1} = (c_x^{f-1}, c_y^{f-1}, w^{f-1}, h^{f-1})$ before non-maximum suppression (NMS). Taking inspiration from NMS, we leverage IoU-based suppression to this end. Referring to Algorithm 5.1, after selection by confidence threshold, IoU between candidate boxes and b^{f-1} is calculated and then candidate boxes with small IoU (e.g., $<U^{sos}$) are discarded. Next, tracking failure would be reported if all boxes are suppressed by SOS. Subsequently, NMS is performed on the remaining boxes. Finally, we select a box with IoU maximum as current SOT result in b^f. Compared to the manner of IoU-based rescoring, the SOS does not affect confidence scores. In our opinion, confidence score and IoU are two different properties of objects, where confidence score describes object category while IoU reports object motion. Thereby, the SOS-NMS is based on alternating confidence score and IoU, i.e., (1) discarding obviously category-incorrect candidates with confidence threshold; (2) discarding obviously motion-incorrect candidates using IoU threshold; (3) discarding candidates without a local maximum of confidence score; and (4) generating single object location with IoU maximum. Note that SOS-NMS has a speed advantage over NMS since a significant amount of candidate boxes are suppressed by computationally efficient SOS.

5.4.2 SOT-by-Detection Framework

As shown in Figure 5.3, the proposed SOT-by-detection framework only adopts a detection model, and an MOT branch is developed to search initial object location for SOT. We first define the condition of MOT-SOT switch: (1) MOT is initially performed; (2) when a reliable tracklet (i.e., $S_{dur} > S_{SDE}$ captured by OTR) is found, the SOT branch is activated to track this reliable tracklet. If there are several reliable tracklets, only one would be tracked according to confidence score; (3) the MOT branch is reactivated after SOT failure captured by SOS. In terms of components, the MOT branch includes NMS, data association, and OTR; the SOT branch contains SOS, NMS, tracklet update, and OTR. For SOT, OTR only processes the tracked tracklet. Remarkably, the SOT branch is faster than the MOT branch because (1) SOS is able to highly reduce NMS's computational cost and (2) data association in the MOT branch is usually time-consuming. The proposed SOT-by-detection has threefold advantages in SOT task: (1) There is no need for man-made initial location; (2) confidence score is more reliable than similarity score because of semantic

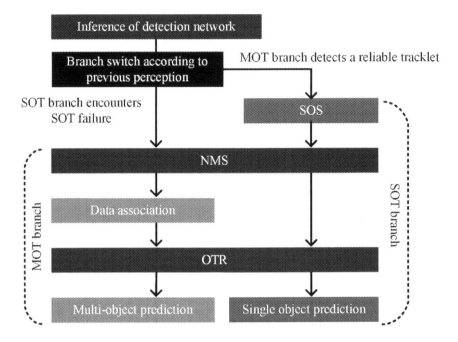

FIGURE 5.3 SOT-by-detection with our proposed SOS and OTR. The MOT branch is designed to search the initial box for SOT. Reliable tracklet is captured by OTR, while SOT failure is captured by SOS. If no switch condition is met, the previous behavior is continuously performed.

classification, so SOT-by-detection can capture tracker state (e.g., tracking drift or failure) in a more reliable manner; (3) complex detection-tracking cascade is avoided. Furthermore, our proposed framework can be easily extended to two-object tracking, three-object tracking, and so on.

5.5 EXPERIMENTS AND DISCUSSION

We analyze real-time online detectors, i.e., SSD [24], RetinaNet [25], RefineDet [26], DRNet [27], TSSD [9], TRNet, and TDRNet [11]. The first four are static detectors while the last three are temporal methods. These detectors have close relation and inheritance, so we unveil the effects of their designs on temporal performance based on our metrics. In return, these evaluations can verify the effectiveness of our assessments. SSD detects objects in a single-stage manner [24]. Based on SSD, RetinaNet adopts feature pyramid networks (FPN) to enhance shadowlayer receptive field [25]. Based on RetinaNet, RefineDet introduces a two-step regression to the single-stage pipeline [26]. Based on RefineDet, DRNet performs joint anchor-feature refinement for detection [27]. Referring to Section 5.2, there are five types of VID approaches, but post-processing and tracking-based methods actually adopt static detectors, and batch-frame approaches can hardly work in real-world scenes, so we analyze the methods with feature aggregation or temporally sustained proposal. Based on SSD, TSSD uses attentional-LSTM for aggregating visual features across time [9]. As temporally sustained proposal approaches, TRNet and TDRNet propagate refined anchors and feature offsets across time based on RefineDet and DRNet [11]. All these detectors are trained and evaluated on ImageNet VID dataset [1]. Both confidence threshold and NMS threshold are fixed as 0.5.

5.5.1 Analysis on VID Continuity/Stability

5.5.1.1 Tracklet Visualization

As shown in Figure 5.4a, we use a VID case with nine object instances to visualize SSD detection. Referring to Figure 5.4b, SSD suffers from serious continuity and stability problems. At the beginning of this video, a vast number of objects missing (i.e., "×" on curves) and short tracklets (i.e., short curves and scattered points) appear due to motion blur. Then, continuity problems appear again at the end of the video owing to occlusion. For localization stability, Figure 5.4d plots the amplitude of high-frequency component (>0.1 Hz) in the Fourier domain. Numerically, ESDE = 0.448, SDE = 0.715, TFE = 0.410, FTR = 0.619, RCE = 2.192, CJE = 0.264, SJE = 0.183, and LJE = 0.447.

OTR is able to refine SSD results from the perspective of tracklet. As shown in Figure 5.4c, OTR eliminates all short tracklets and fragments, and refined tracklets are smoother. Referring to Figure 5.4e, high-frequency amplitude in Fourier domain is suppressed to some degree. As a result, ESDE = 0.0, SDE = 0.0, TFE = 0.0, FTR = 0.9, RCE = 0.0, CJE = 0.219, SJE = 0.142, and LJE = 0.361.

5.5.1.2 Numerical Evaluation

Referring to Table 5.1, detectors are evaluated with our proposed non-reference assessments on VID validation set. The accuracy-best method is DRNet, with 69.4% mAP. However, there is a low correlation between accuracy and continuity/stability. From static SSD and RetinaNet, we observe that FPN improves localization stability since spatial feature fusion. However, as for continuity, RetinaNet performs worse than SSD since more hard objects can be detected by RetinaNet. That is, detecting hard objects (e.g., small

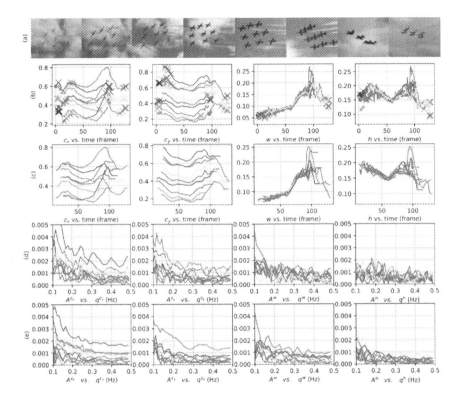

FIGURE 5.4 Tracklet visualization for SSD. (a) The video snippet; (b) original tracklets; (c) refined tracklets with OTR; (d) original Fourier results; (e) Fourier results with OTR. Colors differentiate ID and "×" denotes object missing at a timestamp. A and q are defined in Section 5.3.1.

objects) can easily produce continuity problems, and SSD is likely to completely miss them because of relatively low detection accuracy. Besides, the comparison between RefineDet and RetinaNet indicates that anchor refinement improves accuracy but induces more serious problems on continuity and stability because anchor-feature misalignment is exacerbated. Finally, DRNet conducts joint anchor-feature refinement to relieve RefineDet's drawback, i.e., features are relatively accurate for describing refined anchors. Thus, DRNet performs better than RefineDet on almost all metrics.

Based on OTR, all metrics can be effectively improved for all tested approaches. OTR can totally eliminate fragment problem by filling up recall failure, generating 0 TFE/FTR. In terms of ESDE, SDE, CJE, and SJE, OTR also produces substantially better results. Note that AP cannot be reported with OTR, because MOT needs a relatively high confidence

TABLE 5.1 Continuity and stability evaluation of several existing detectors based on the proposed non-reference metrics

Method	mAP	Recall continuity					Localization stability ($\omega = 10$)		
		ESDE	SDE	TFE	FTR	RCE	CJE	SJE	LJE
w/o OTR									
static method									
SSD [24]	0.630	0.062	0.234	0.320	0.246	0.862	0.242	0.334	0.576
RetinaNet [25]	0.656	0.060	0.250	0.350	0.283	0.943	0.236	0.317	0.553
RefineDet [26]	0.669	0.126	0.350	0.391	0.306	1.173	0.257	0.362	0.619
DRNet [27]	**0.694**	0.114	0.330	0.389	0.312	1.145	0.248	0.346	0.594
Temporal method									
TRNet [13]	0.665	0.120	0.334	0.375	0.265	1.094	0.252	0.346	0.598
TDRNet [13]	0.673	0.116	0.345	0.388	0.297	1.146	0.247	0.360	0.607
TSSD [11]	0.654	**0.059**	**0.206**	**0.257**	**0.240**	**0.762**	**0.210**	**0.253**	**0.463**
w/OTR									
SSD	–	0.003	0.026	0.0	0.0	0.029	0.169	0.208	0.377
RetinaNet	–	0.003	0.023	0.0	0.0	0.026	0.168	0.204	0.372
RefineDet	–	0.004	0.037	0.0	0.0	0.041	0.173	0.212	0.385
DRNet	–	0.003	0.036	0.0	0.0	0.039	0.172	0.208	0.380
TRNet	–	0.003	0.030	0.0	0.0	0.033	0.171	0.209	0.380
TDRNet	–	0.004	0.031	0.0	0.0	0.035	0.170	0.218	0.388
TSSD	–	0.003	0.029	0.0	0.0	0.032	0.159	0.180	0.339

threshold (e.g., 0.5) for data association while AP evaluation usually uses a low threshold (i.e., 0.01).

We use $\Omega = \{\omega^l \,|\, l = 1,\ldots,0.1\}$ to fuse current prediction with location history, so Ω controls the temporal location fusion. We investigate $\omega = 1,2,5,10,20,40$, which induces decay ratios of $1,1.17,1.44,1.68,1.96,2.29$ across time. That is, small ω merges location information with roughly equal weights, while large Ω produces greater weights for recent timestamps. Note that $\omega = \infty$ means temporal location fusion loses its effect. For example, $\Omega = \{0.437,0.260,0.155,0.092,0.055\}$, when $\omega = 10$, $S_{obj} = 5$. Referring to Figure 5.5, optimal ω ranges from 5 to 10, where the fusion ratio is suitable for localization stability.

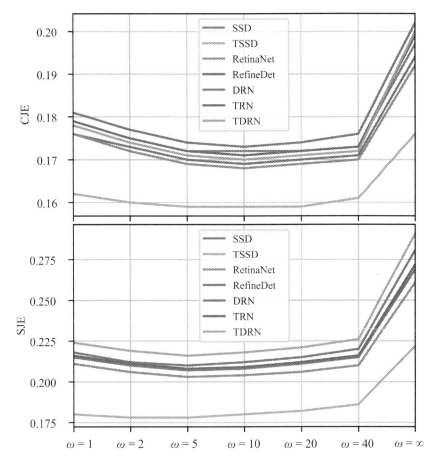

FIGURE 5.5 Plot of CJE/SJE vs. ω. $\omega = \infty$ indicates that temporal location fusion loses its effect.

5.5.2 SOT-by-Detection

5.5.2.1 Speed Comparison of NMS and SOS-NMS

This test is conducted with an Intel 2.20 GHz Xeon(R) E5-2630 CPU. As plotted in Figure 5.6, the SOS considerably reduces the time consumption of candidate selection. NMS takes 2.30 ms per frame, and SOS can highly reduce the box amount for NMS with temporal information. As a result, SOS-NMS's time cost can be reduced to 0.69 ms per frame, and the NMS part in SOS-NMS only takes 0.33 ms per frame when SOS threshold >0.3. Therefore, when performing SOT based on SOS, the VID model can achieve faster speed. In the following experiments, SOS threshold is fixed as 0.3.

5.5.2.2 SOT-by-Detection vs. Siamese SOT

There lacks a dataset to quantitatively evaluate SOT-by-detection and Siamese SOT. That is, Siamese SOT cannot work on VID dataset (e.g., ImageNet VID [1]) because it cannot predict object category, and VID model in SOT-by-detection cannot be trained with a category-free SOT dataset (e.g., VOT [13]). Thus, we qualitatively compare our method and a Siamese tracker [16] on ImageNet VID dataset. Firstly, SOT-by-detection has the ability to search objects, i.e., MOT can provide an initial localization for SOT. Then, Siamese SOT is particularly susceptible to unconscious tracking drift (as delineated in Figure 5.7a). Finally, referring to Figure 5.7b, SOT-by-detection is able to capture tracking failure and conveniently reactivate the MOT branch for object search (see the third snapshot). On the contrary, the Siamese tracker always reports a similar region, and it is unable to start a second tracking.

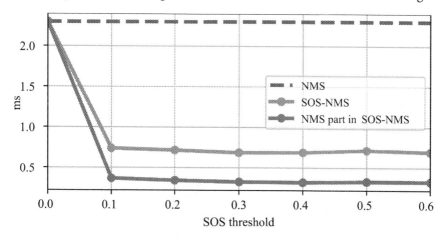

FIGURE 5.6   Plot of time consumption of NMS and SOS-NMS.

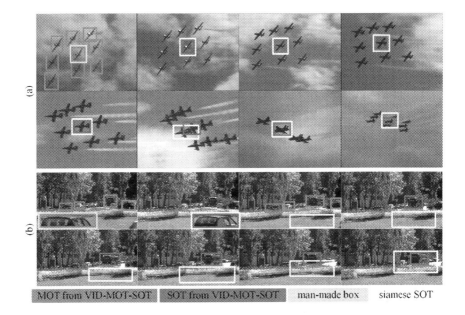

FIGURE 5.7 Comparison between SOT-by-detection and Siamese SOT. If two boxes are highly overlapping, we slightly change their sizes for visualization.

5.5.3 Discussion

5.5.3.1 Detector-Based Improvement

This chapter evaluates and enhances detection continuity and stability from the perspective of MOT. Additionally, our evaluation indicates that temporal performance is benefited from feature aggregation. On the one hand, spatial/scale smooth is effective (see RetinaNet vs. SSD), and on the other hand, temporal fusion is more efficient (see TSSD vs. SSD). Thus, we advocate investigating fusion approaches for improving continuity/stability from the detector itself. For example, besides RNN in TSSD, Zhu et al. aggregated temporal features with flow-based motion estimation [7], and Bertasius et al. leveraged the deformable convolutions to constructed robust temporal features [8].

5.5.3.2 Limitation of SOT-by-Detection

SOT-by-detection has advantages, but it also has a limitation that it cannot deal with unseen categories or object part. On the one hand, similarity-based SOT and SOT-by-detection are complementary. That is, the similarity module focuses on appearance similarity while SOT-by-detection is dependent on category attribute and motion information. Therefore, they can be reasonably combined for addressing this problem. On the other hand, we advocate solving this limitation with a combination of online learning [28] and few short learning [29].

5.6 CONCLUDING REMARKS

This chapter has dealt with detection continuity/stability and SOT-by-detection for robotic perception. Firstly, we analyze that AP has difficulty in reflecting temporal continuity/stability and propose non-reference assessments to evaluate them. The evaluation of recall continuity is based on tracklet analysis, and the localization stability problem is captured by the Fourier method. Secondly, we design OTR to enhance detection continuity/stability by short tracklet suppression, fragment filling, and temporal location fusion. Finally, SOS is proposed to extend VID methods toward SOT task, and the SOT-by-detection is developed. With ImageNet VID dataset, the effects of our methods are verified. As a result, our non-reference assessments and OTR can effectively deal with temporal performance in VID, and SOT-by-detection plays an essential role in robotic perception.

We will enhance temporal performance with feature fusion and improve SOT-by-detection by online learning.

REFERENCES

1. Russakovsky, J. Deng, H. Su, J. Krause, S. Satheesh, S. Ma, Z. Huang, A. Karpathy, A. Khosla, M. Bernstein, A. C. Berg, and F. Li, "ImageNet large scale visual recognition challenge," *Int. J. Comput. Vision*, vol. 115, no. 3, pp. 211–252, 2015.
2. K. Kang, W. Ouyang, H. Li, X. Wang, "Object detection from video tubelets with convolutional neural networks," in *Proceedings of the IEEE Conference on Computer Vision and Pattern Recognition*, Las Vegas, USA, Jun. 2016, pp. 817–825.
3. C. Feichtenhofer, A. Pinz, A. Zisserman, "Detect to track and track to detect," in *Proceedings of the IEEE Conference on Computer Vision and Pattern Recognition*, Venice, Italy, Oct. 2017, pp. 3038–3046.
4. W. Han, P. Khorrami, T. L. Paine, P. Ramachandran, M. Babaeizadeh, H. Shi, J. Li, S. Yan, and T. S. Huang, "Seq-NMS for video object detection," https://arxiv.org/abs/1602.08465, 2016.

5. H. Luo, W. Xie, X. Wang, W. Zeng, "Detect or track: Towards cost effective video object detection/tracking," in *Proceedings of the AAAI Conference on Artificial and Intelligence*, Honolulu, USA, Jul. 2019, pp. 8803–8810.

6. K. Kang, H. Li, J. Yan, X. Zeng, B. Yang, T. Xiao, C. Zhang, Z. Wang, R. Wang, X. Wang, and W. Ouyang, "T-CNN: Tubelets with convolutional neural networks for object detection from videos," *IEEE Trans. Circuits Syst. Video Technol.*, vol. 28, no. 10, pp. 2896–2907.

7. X. Zhu, J. Dai, L. Yuan, and Y. Wei, "Towards high performance video object detection," in *Proceedings of the IEEE Conference on Computer Vision and Pattern Recognition*, Salt Lake City, USA, Jun. 2018, pp. 7210–7218.

8. G. Bertasius, L. Torresani, and J. Shi, "Object detection in video with spatiotemporal sampling networks," in *Proceedings of the European Conference on Computer Vision*, Munich, Germany, Sept. 2018, pp. 342–357.

9. X. Chen, J. Yu, and Z. Wu, "Temporally identity-aware SSD with attentional LSTM," *IEEE Trans. Cybern.*, vol. 50, no. 6, pp. 2674–2686, 2020.

10. K. Kang, H. Li, T. Xiao, W. Ouyang, J. Yan, X. Liu, and X. Wang, "Object detection in videos with tubelet proposal networks," in *Proceedings of the IEEE Conference on Computer Vision and Pattern Recognition*, Hawaii, USA, Jul, 2017, pp. 727–735.

11. X. Chen, J. Yu, S. Kong, Z. Wu, and L. Wen, "Joint anchor-feature refinement for real-time accurate object detection in images and videos," *IEEE Trans. Circuits Syst. Video Technol.*, in press, doi:10.1109/TCSVT.2020.2980876. https://ieeexplore.ieee.org/abstract/document/9037090/

12. L. Leal-Taixe, A. Milan, I. Reid, S. Roth, and K. Schindler, "Motchallenge 2015: Towards a benchmark for multi-target tracking," https://arxiv.org/abs/1504.01942, 2015.

13. M. Kristan, et al., "The sixth visual object tracking VOT-2018 challenge results," in *Proceedings of the European Conference on Computer Vision Workshops*, Munich, Germany, Sept. 2018.

14. J. F. Henriques, R. Caseiro, P. Martins, J. Batista, "High-speed tracking with kernelized correlation filters," *IEEE Trans. Pattern Anal. Mach. Intell.*, vol. 37, no. 3, pp. 583–596, 2014.

15. B. Li, J. Yan, W. Wu, Z. Zhu, X. Hu, "High performance visual tracking with siamese region proposal network," in *Proceedings of the IEEE Conference on Computer Vision and Pattern Recognition*, Salt Lake City, USA, Jun. 2018, pp. 8971–8980.

16. Y. Xu, Z. Wang, Z. Li, Y. Yuan, and G. Yu, "SiamFC++: Towards robust and accurate visual tracking with target estimation guidelines," in *Proc. AAAI Conference on Artificial Intelligence*, New York, USA, Feb. 2020, pp. 12549–12556. *arXiv:1911.06188*, 2019.

17. L. Pang, Z. Cao, J. Yu, P. Guan, X. Rong, H. Chai, "A visual leader-following approach with a TDR framework for quadruped robots," *IEEE Trans. on Syst. Man Cybern. Syst.*, 2019, doi: 10.1109/TSMC.2019.2912715. https://ieeexplore.ieee.org/abstract/document/8709995

18. H. U. Kim and C. S. Kim, "CDT: Cooperative detection and tracking for tracing multiple objects in video sequences," in *Proceedings of the European Conference on Computer Vision*, Amsterdam, Netherlands, Oct. 2016, pp. 851–867.
19. M. Everingham, L. Van Gool, C. K. Williams, J. Winn, and A. Zisserman, "The pascal visual object classes (VOC) challenge," *Int. J. Comput. Vision*, vol. 88, no. 2, pp. 303–338, 2010.
20. H. Zhang, N. Wang, "On the stability of video detection and tracking," https://arxiv.org/abs/1611.06467, 2016.
21. K. Bernardin and R. Stiefelhagen, "Evaluating multiple object tracking performance: The CLEAR MOT metrics," *EURASIP J. Image Video Process.*, vol. 1, pp. 1–10, 2008.
22. Bewley, Z. Ge, L. Ott, F. Ramos, and B. Upcroft, "Simple online and realtime tracking," in *IEEE International Conference on Image Processing*, Phoneix, USA, Sep. 2016, pp. 3464–3468.
23. Y. Lu, C. Lu, and C. K. Tang, "Online video object detection using association LSTM," in *Proceedings of the IEEE International Conference on Computer Vision*, Venice, Italy, Oct. 2017, pp. 2344–2352.
24. W. Liu, D. Anguelov, D. Erhan, C. Szegedy, S. Reed, C. Y. Fu, and A. C. Berg, "SSD: Single shot multibox detector," in *Proceedings of the European Conference on Computer Vision*, Amsterdam, Netherlands, Oct. 2016, pp. 21–37.
25. T. Y. Lin, P. Goyal, R. Girshick, K. He, and P. Dollar, "Focal loss for dense object detection," in *Proceedings of the IEEE International Conference on Computer Vision*, Venice, Italy, Oct. 2017, pp. 2980–2988.
26. S. Zhang, L. Wen, X. Bian, Z. Lei, S. Z. Li, "Single-shot refinement neural network for object detection," in *Proceedings of the IEEE International Conference on Computer Vision and Pattern Recognition*, Salt Lake City, USA, Jun. 2018, pp. 4203–4212.
27. X. Chen, X. Yang, S. Kong Z. Wu, and J. Yu, "Dual refinement network for single-shot object detection," in *Proceedings of the International Conference on Robotics and Automation*, Montreal, Canada, May 2019, pp. 8305–8310.
28. M. Danelljan, G. Bhat, F. S. Khan, M. Felsberg, "Atom: Accurate tracking by overlap maximization," in *Proceedings of the IEEE Conference on Computer Vision and Pattern Recognition*, Long Beach, USA, Jun. 2019, pp. 4660–4669.
29. Li, T. Luo, Z. Lu, T. Xiang, and L. Wang, "Large-scale few-shot learning: Knowledge transfer with class hierarchy," in *Proceedings of the IEEE International Conference on Computer Vision and Pattern Recognition*, Long Beach, USA, Jun. 2019, pp. 7212–7220.

CHAPTER 6

Reveal of Domain Effect

How Visual Restoration Contributes to Object Detection in Aquatic Scenes

6.1 INTRODUCTION

Within the past few years, great efforts have been made for underwater robotics. For example, Gong et al. designed a soft robotic arm for underwater operation [1]. Cai et al. developed a hybrid-driven underwater vehicle manipulator for collecting marine products [2]. Toward intelligent autonomous robots, visual methods are usually adopted for underwater scene perception [1–4].

With the advent of convolutional neural network (CNN), object detection has been a surging topic in computer vision [5–8], and in the meantime, object detection is a fundamental tactic for robotic perception [9]. Based on the detection, robots can discover what and where the target is. However, because of optical absorption and scattering, the underwater visual signal usually suffers from degeneration and forms low-quality images/videos [10]. Note that low quality means low contrast, high color distortion, and strong haziness. Therefore, visual restoration has been widely studied [10–15] so that visual quality can be improved for subsequent image processing. By and large, visual restoration and object detection are two essential abilities for an aquatic robot to perform object perception.

Although visual restoration has been proven to be helpful for traditional man-made features (e.g., SIFT [16] [12], the relation between image quality and convolutional representation remains unclear. As demonstrated in Figure 6.1, underwater scenes are always degenerated, and moreover, the degeneration usually has different styles, i.e., color distortion, haziness, and illumination (see the top line). By filtering-based restoration (FRS) [14] and GAN-based restoration (GAN-RS) [15], higher-quality images are generated. Although each column of Figure 6.1 is the same scenario, their detection results are diverse with DRNet detector [8]. Therefore, scopes of restoration and detection should have latent relevance that should be

original underwater scenes

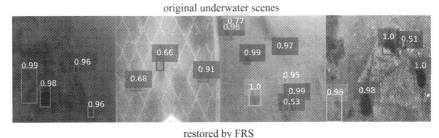

restored by FRS

restored by GAN-RS

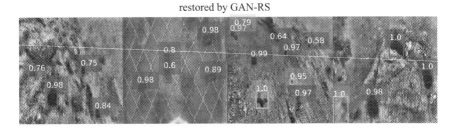

FIGURE 6.1 Underwater object detection based on different restoration manners. Underwater visual degeneration is diverse, as shown in the top line. Relieving this degeneration, FRS [14] and GAN-RS [15] generate clear images. Further, for the same scenario, different detection results are produced because of different restoration methods.

investigated. To this end, we study to answer a question – *How does visual restoration contribute to object detection in aquatic scenes?*

In addition, visual restoration exactly produces the change of data domain, and it is known that the data domain is important for data-driven learning process [17–21]. However, under the condition of different data domains, within-domain and cross-domain detection performances have rarely been studied. That is, the domain effect on object detection remains unclear. In our opinion, exploring the effect of the data domain is instructive for building robust real-world detectors. Thereby, we are motivated to investigate the relation between image quality and detection performance based on visual restoration to unveil domain effect on object detection. In this way, the relation of restoration to detection can also be exposed. In this chapter, we joint analyze visual restoration and object detection for underwater robotic perception. At first, we construct quality-diverse data domains with FRS and GAN-RS for both training and testing. FRS is a traditional filtering method, and GAN-RS is a learning-based scheme so that they can be representative for the restoration sphere. Further, we investigate typical single-stage detectors (i.e., SSD [5], RetinaNet [7], RefineDet [6], and DRNet [8]) on different data domains and then within-domain and cross-domain performances are analyzed. Finally, real-world experiments are conducted on the seabed for online object detection. Based on our study, the relation of restoration-based data domain to detection performance is unveiled. As a result, although it induces adverse effects on object detection, visual restoration efficiently suppresses domain shift (i.e., discordance between training domain and testing domain) between training images and practical scenes. Thus, visual restoration still plays an essential role in aquatic robotic perception. Our contributions are summarized as follows:

- We reveal three-domain effects on detection: (1) Domain quality has a negligible effect on within-domain convolutional representation and detection accuracy after sufficient training; (2) low-quality domain brings about better generalization in cross-domain detection; (3) in domain-mixed training, a low-quality domain can hardly be well learned.

- We indicate that restoration is a thankless operation for improving within-domain detection accuracy. In detail, it reduces recall efficiency [22]. However, visual restoration is beneficial in reducing domain shift between training data and practical aquatic scenes so that online detection performance can be boosted. Therefore, it is an essential operation in real-world object perception.

- Based on our analysis, online object detection is successfully conducted on the field unstructured seabed with an aquatic vision-based robot.

6.2 REVIEW OF UNDERWATER VISUAL RESTORATION AND DOMAIN-ADAPTIVE OBJECT DETECTION

6.2.1 Underwater Visual Restoration

Because of natural physical phenomenon, the underwater visual signal is usually degenerated, forming low-quality vision. In detail, underwater image/video shows low contrast, high color distortion, and strong haziness, making image processing difficult. Schechner and Karpel attributed this degeneration to visual absorption and scattering [10]. Overcoming this difficulty, Peng and Cosman proposed a restoration method based on image blurriness and light absorption, which estimated scene depth for image formation model [13]. Chen et al. adopted filtering model and artificial fish algorithm for real-time visual restoration [14]. Li et al. hierarchically estimated background light and transmission map, and their method was characterized by minimum information loss [12]. Chen et al. proposed a weakly supervised GAN and an adversarial critic training to achieve real-time adaptive restoration [15]. Recently, Liu et al. built an underwater enhancement benchmark for follow-up works, whose samples were collected on the seabed under natural light [11].

With the abovementioned studies, it is revealed that visual restoration is beneficial in clearing image details and producing salient low-level features. For example, canonical SIFT [16] algorithms deliver a huge performance improvement based on restoration [12]. However, how visual restoration contributes to CNN-based feature representation remains unclear. Moreover, visual restoration is tightly related to the data domain, so we explore domain effect based on restoration.

6.2.2 Domain-Adaptive Object Detection

During the deep learning era, single-stage object detection uses a single-shot network for regression and classification. As a pioneering work, Liu et al. proposed SSD for real-time detection [5]. Inspired by a feature pyramid network, Li et al. developed RetinaNet to propagate CNN features in a top-down manner for enlarging shallow layers' receptive field [7]. Zhang et al. introduced two-step regression to the single-stage pipeline and designed RefineDet for addressing the class imbalance problem. Chen et al. proposed DRNet with anchor-offset detection that achieved single-stage region proposal [8]. Although some two-stage detectors [23] and anchor-free detectors [24] could induce higher accuracy, the single-stage methods maintain a better accuracy-speed trade-off for robotic tasks.

The above detectors generally assume that training and test samples fall within an identical distribution. However, real-world data usually suffer from a domain shift, which affects detection performance. Hence, cross-domain robustness of object detection is recently explored. Chen et al. proposed adaptive components for image-level and instance-level domain shift based on H-divergence theory [17]. Xu et al. utilized a deformable part-based model and adaptive SVM for mitigating domain shift problem [18]. Raj et al. developed a subspace alignment approach for detecting an object in real-world scenarios [19]. For alleviating the problem of domain shift, Khodabandeh et al. exploited a robust learning method with noisy labels [20]. Inoue et al. proposed a cross-domain weakly supervised training based on domain transfer and pseudo-labeling for domain-adaptive object detection [21].

These works have indicated how to moderate the domain shift problem, but there has been relatively little work extensively studying the domain effect on detection performance. In contrast, based on underwater scenarios, we investigate the effect of quality-diverse data domain on object detection. Kalogeiton et al. analyzed detection performance based on different image quality [25], but we have advantages over their work: (1) [25] was reported before deep learning era, but we analyze deep learning-based object detection; (2) [25] considered the impact of simple factors (e.g., Gaussian blur), but our domain change is derived from realistic visual restoration; (3) [25] only analyzed cross-domain performance, but we investigate both cross-domain and within-domain performances; and (4) our work contributes to aquatic robotics.

6.3 PRELIMINARY

6.3.1 Preliminary of Data Domain Based on Visual Restoration

6.3.1.1 Domain Generation

The dataset is publicly available for underwater object detection, i.e., Underwater Robotic Picking Contest 2018 (URPC2018). This dataset is collected on the natural seabed at Zhangzidao, Dalian, China. URPC2018 is composed of 2,901 aquatic images for training and 800 samples for testing. In addition, it contains four categories, i.e., "trepang", "echinus", "shell", and "starfish".

Based on URPC2018, three data domains are generated: (1) *domain-O*: The original dataset with *train* set and *test* set; (2) *domain-F*: All samples are processed by FRS, producing *train-F* set for training and *test-F* set for testing; (3) *domain-G*: All samples are restored by GAN-RS, generating *train-G* set for training and *test-G* set for testing. Mixed *train*, *train-F*, and *train-G* are denoted as train-all. As shown in Figure 6.2, *domain-O* has strong color distortion, haziness, and low contrast. The degenerated visual samples are effectively restored in *domain-F* and *domain-G*.

6.3.1.2 Domain Analysis

According to [15], Lab color space has good ability to describe underwater properties of images. Thus, Figure 6.3 illustrates a–b distribution in Lab color space. As a result, the distribution of *domain-O* consistently gathers far from the color balance point (i.e., (128,128)). The bias between distribution center and the balance point means strong color distortion, and the concentrated distribution indicates strong haziness. On the contrary, different from *domain-O*, the distributions of *domain-F* and *domain-G* have a trend of color balance and haze removal.

Underwater color image quality evaluation metric (UCIQE) [26] and underwater image quality measures (UICM, UISM, UIConM, UIQM) [27] are used to describe domain quality. UCIQE quantifies image quality via chrominance, saturation, and contrast. UIQM is a comprehensive quality representation of an underwater image, in which UICM, UISM, and UIConM separately describe color, sharpness, and contrast. Referring to Table 6.1, benefited from visual restoration, *domain-F* brings about best UCIQE and UICM while *domain-G* induces the best UISM, UIConM, and UIQM. Therefore, we define *domain-F* and *domain-G* as high-quality domains with high-quality samples. In contrast, *domain-O* is defined as a low-quality domain with low-quality samples. Besides, referring to Figure 6.3 and Table 6.1, GAN-RS has better restoration results, so we define that GAN-RS induces a higher restoration intensity than FRS.

6.3.2 Preliminary of Detector

According to [28], two-stage methods have no advantage over single-stage approaches on URPC2018. Therefore, because of the ability to induce both

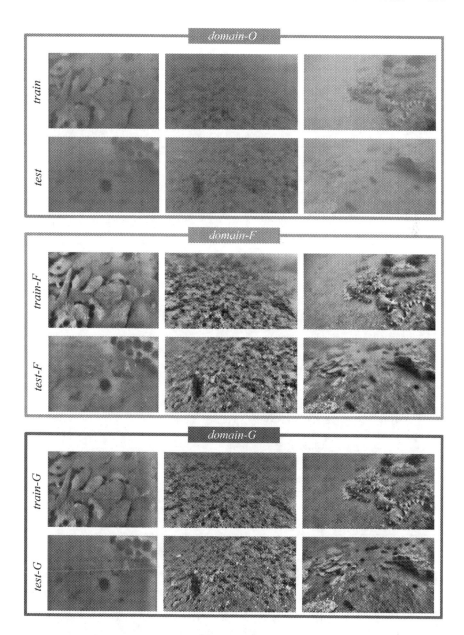

FIGURE 6.2   Typical samples in *domain-O*, *domain-F*, and *domain-G*.

high accuracy and real-time inference speed, we leveraged single-stage detectors to perform underwater offline/online object detection. In detail, this chapter investigates SSD, RetinaNet, RefineDet, and DRNet. All these detectors are trained based on *train, train-F, train-G,* or train-all. As for training details, an SGD optimizer with 0.9 momentum and 5×10^{-4}

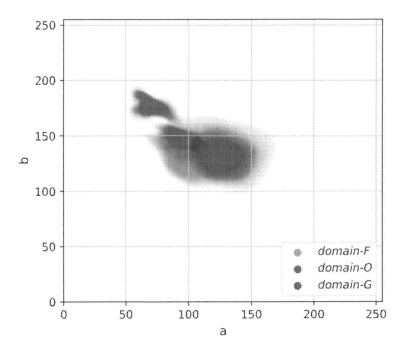

FIGURE 6.3 Domain visualization in Lab color space. a–b distribution of *domain-O* is concentrated and has a color bias. In contrast, distributions of *domain-F* and *domain-G* are more scattered and have smaller biases. Color transparency indicates distribution probability.

TABLE 6.1   Quality assessment for data domain

Domain	UCIQE	UICM	UISM	UIConM	UIQM
domain-O	0.39	0.20	3.86	0.12	1.58
domain-F	**0.56**	**3.38**	12.87	0.17	4.51
domain-G	0.53	2.27	**13.81**	**0.18**	**4.78**

weight decay is employed, and batch size is 32. We use the initial learning rate of 10^{-3} for the first 12×10^3 iteration steps, then use the learning rate of 10^{-4} for the next 3×10^3 steps and 10^{-5} for another 3×10^3 steps. In this manner, all detectors can be sufficiently trained. For evaluation, mean average precision (mAP) is employed to describe detection accuracy.

6.4 JOINT ANALYSIS ON VISUAL RESTORATION AND OBJECT DETECTION

6.4.1 Within-Domain Performance

In this test, detectors' training and evaluation are based on identical data domain. The following analysis will unveil two points: (1) Domain quality has an ignorable effect on detection performance; (2) restoration is a thankless method for improving within-domain detection performance, because of the problem of low recall efficiency. Note that low recall efficiency means low precision under the condition of the same recall rate [22].

6.4.1.1 Numerical Analysis

At first, we train and evaluate SSD with different input sizes (i.e., 320 and 512) and backbones (i.e., VGG16 [29], MobileNet [30], and ResNet101 [31]). As shown in Table 6.2, on *domain-O*, *domain-F*, and *domain-G*, SSD320-VGG16 achieves mAP of 69.3%, 67.8%, and 65.9% and SSD512-VGG16 obtains mAP of 72.9%, 71.3%, and 69.5%. It is seen that the accuracy decreases with the rise of restoration intensity. From backbone-variable assessments, the same trend emerges. Note that ResNet101 performs inferiorly to VGG16 and MobileNet, because the large receptive field in ResNet101 is unfavorable to an immense number of small objects in URPC2018. Referring to Table 6.2, all of RetinaNet512, RefineDet512, and DRNet512 can achieve the highest mAP on *domain-O* and see the lowest mAP on *domain-G*. Thus, in terms of mAP, detection accuracy is negatively correlated with domain quality. However, mAP cannot reflect accuracy details, so the following analysis will continue investigating within-domain performance.

6.4.1.2 Visualization of Convolutional Representation

The human perceives domain quality based on object saliency. As a result, compared to the low-quality domain, the human can more easily detect objects in the high-quality domain since high-quality samples contain salient object representation. Thereby, we are inspired to investigate object saliency in CNN-based detectors. Figure 6.4 demonstrates multi-scale features in SSD and DRNet. These features serve as the input of detection heads, so they are final convolutional features for detection. Referring to Figure 6.4, despite domain diversity, there is relatively little difference in object saliency in multi-scale feature maps. It is seen that different from a human's perception mechanism, convolution is able to capture salient object representation from low-quality data domain. Hence, in terms of object saliency, domain quality has an ignorable effect on convolutional representation.

6.4.1.3 Precision-Recall Analysis

As shown in Figure 6.5, precision-recall curves are employed for further analysis of detection performance. It can be seen that precision-recall

TABLE 6.2 Within-domain detection results

Method	Train data	Test data	mAP	trepang	echinus	shell	starfish
SSD320-VGG16	train	test	**69.3**	**67.8**	**84.9**	**44.7**	**79.7**
	train-F	test-F	67.8	68.9	82.3	42.2	78.0
	train-G	test-G	65.9	65.4	82.3	39.0	76.9
SSD512-VGG16	train	test	**72.9**	**70.2**	**87.1**	**50.8**	**83.5**
	train-F	test-F	71.3	68.9	85.8	48.5	82.1
	train-G	test-G	69.5	67.2	84.7	45.3	80.9
SSD512-MobileNet	train	test	**70.7**	**65.3**	**87.1**	**47.5**	**82.8**
	train-F	test-F	68.9	63.7	85.1	45.4	81.7
	train-G	test-G	67.4	61.5	84.9	42.6	80.5
SSD512-ResNet101	train	test	**67.0**	59.8	**86.3**	**41.7**	**80.3**
	train-F	test-F	65.6	**61.1**	84.7	37.5	79.1
	train-G	test-G	64.6	60.1	83.7	38.6	76.2
RetinaNet512-VGG16	train	test	**74.0**	**69.8**	**88.1**	**54.7**	**83.4**
	train-F	test-F	72.5	69.1	87.1	50.7	82.9
	train-G	test-G	71.0	67.3	86.9	48.9	81.1
RefineDet512-VGG16	train	test	**76.0**	**73.8**	**90.2**	**54.1**	**85.8**
	train-F	test-F	72.9	72.0	88.6	46.4	84.6
	train-G	test-G	72.0	71.4	88.4	46.3	81.8
DRNet512-VGG16	train	test	**77.1**	**75.6**	**91.1**	**55.1**	**86.7**
	train-F	test-F	75.4	73.6	89.8	52.7	85.6
	train-G	test-G	73.8	72.0	89.8	49.9	83.5

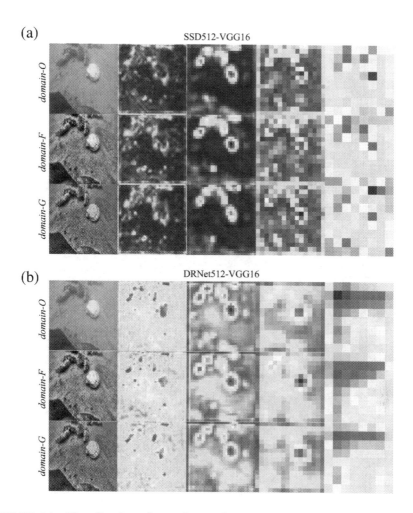

FIGURE 6.4   Visualization of convolutional representation for objects. Each row contains input image and multi-scale features. High-level features are shown on the right. All features are processed with L2 norm across channel and then they are normalized for visualization. For a fair comparison, the same normalization factor is used for scale-identical features.

curves have two typical appearances. On the one hand, the high-precision part contains high-confident detection results, and here domain-related curves are highly overlapped. Referring to "echinus" detected by DRNet512-VGG16, curves of *domain-O*, *domain-F*, and *domain-G* cannot be separated when the recall rate is less than 0.6. That is, when detecting high-confident objects, domain difference is negligible for detection accuracy. On the other hand, curves are separated in the low-precision part.

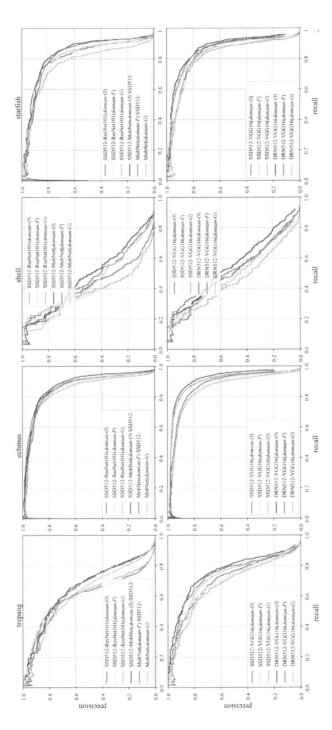

FIGURE 6.5   Precision-recall curves. For high precision (e.g., >0.9), domain difference has an ignorable effect on detection performance. Overall, *domain-F* and *domain-G* reduce recall efficiency so that lower average precision is induced.

In detail, the curve of *domain-F* is usually below that of *domain-O*, while the curve of *domain-G* is usually below that of *domain-F*. That is, when detecting hard objects (i.e., low-confident detection results), false positive increases with the rise of domain quality. For example, when the recall rate equals 0.8 in "starfish" detected by SSD512-VGG16, the precision of *domain-F* is lower than that of *domain-O*, and the precision of *domain-G* is lower than that of *domain-F*. Therefore, recall efficiency is gradually reduced with increasing restoration intensity.

Based on aforementioned analysis, it can be concluded that visual restoration impairs recall efficiency and is unfavorable for improving within-domain detection. In addition, because domain-related mAP is relatively close and high-confident recall is far more important than low-confident recall in robotic perception, we conclude that domain quality has an ignorable effect on within-domain object detection.

6.4.2 Cross-Domain Performance

In this test, detectors are trained and evaluated on different data domains. The following analysis will expose three viewpoints: (1) It is widely accepted that domain shift induces significant accuracy drop; (2) for cross-domain inference, learning based on low-quality domain has better generalization ability toward high-quality domain; (3) in domain-mixed learning, the low-quality domain has smaller contribution so that low-quality samples cannot be well learned.

6.4.2.1 Cross-Domain Evaluation

We use *domain-O* and *domain-G* for evaluation of direction-related domain shift. That is, we train detectors on a train and evaluate them on *test-G*, or vice versa. As shown in Table 6.3, mAP of all categories seriously declines. As a result, if train and *test-G* are employed, SSD512-VGG16 suffers 17.4% mAP drop while DRNet512-VGG16 encounters 15.9% decrease in mAP. However, if *train-G* and test are adopted, SSD and DRNet would suffer from a more dramatic accuracy exacerbation, i.e., mAP drops of 49.4% and 56.3%. According to different degrees of accuracy drop caused by direction-opposite domain shift, it is seen that the generalization of the train toward *test-G* is better than that of *train-G* toward the test. Therefore, it can be concluded that compared to the high-quality domain, the low-quality domain induces better cross-domain generalization ability.

TABLE 6.3   Cross-domain evaluation. "↓" is with respect to within-domain performance of the same test set

Method	Train data	Test data	mAP	trepang	echinus	shell	starfish
SSD512-VGG16	train	test-G	52.1	42.5	70.2	36.6	59.0
			↓ 17.4	↓ 24.7	↓ 14.5	↓ 8.7	↓ 21.9
	train-G	test	23.5	15.5	42.3	12.9	23.3
			↓ 49.4	↓ 54.7	↓ 44.8	↓ 37.9	↓ 60.2
DRNet512-VGG16	train	test-G	57.9	53.7	74.2	40.0	63.7
			↓ 15.9	↓ 18.3	↓ 15.6	↓ 9.9	↓ 19.8
	train-G	test	20.8	7.5	44.5	13.6	17.3
			↓ 56.3	↓ 68.1	↓ 46.6	↓ 41.5	↓ 69.4

TABLE 6.4   Cross-domain training. "↓" and "↑" are with respect to within-domain performance of the same test set

Method	Train data	Test data	mAP	trepang	echinus	shell	starfish
SSD512-VGG16	train-all	test	51.0	34.5	75.6	40.9	53.1
			↓ 21.9	↓ 35.7	↓ 11.5	↓ 9.9	↓ 30.4
		test-F	71.4	69.2	85.4	48.4	82.4
			↑ 0.1	↑ 0.3	↓ 0.4	↓ 0.1	↑ 0.3
		test-G	67.3	63.8	83.0	45.5	76.9
			↓ 2.2	↓ 3.4	↓ 1.7	↑ 0.2	↓ 0.4
DRNet512-VGG16	train-all	test	52.0	34.5	75.6	40.9	53.1
			↓ 25.1	↓ 41.1	↓ 15.5	↓ 14.2	↓ 33.6
		test-F	75.8	75.0	89.8	53.1	85.3
			↑ 0.4	↑ 1.4	0	↑ 0.4	↓ 0.3
		test-G	72.2	70.5	86.6	51.1	80.7
			↓ 1.6	↓ 1.5	↓ 3.2	↑ 1.2	↓ 2.8

6.4.2.2 Cross-Domain Training

For exploring detection performance with domain-mixed learning, we use train-all to train detectors then evaluate them on *test*, *test-F*, and *test-G*. Referring to Table 6.4, on *test-F* and *test-G*, SSD512-VGG16 and DRNet512-VGG16 perform on par with their within-domain performances. However, both SSD512-VGG16 and DRNet512-VGG16 see dramatically worse accuracies on the test, i.e., >20% mAP drop. With the same training settings, within-domain performances can be similarly produced on high-quality *domain-F* and *domain-G*, but low-quality *domain-O* suffers from significant accuracy decline. That is, when train-all is adopted, samples in train lose their effects to some extent. Thus, we conclude that cross-domain training is thankless for improving detection performance. Moreover, quality-diverse data domain has different contributions to the learning process so that low-quality samples cannot be well learned if mixed with high-quality samples.

6.4.3 Domain Effect on Real-World Object Detection

We collect real-world data, namely, online data, for this test. The online data is collected on the natural seabed, located at Jinshitan, Dalian, China. The following analysis will answer the question – *How does visual restoration contribute to object detection?*

6.4.3.1 Online Object Detection in Aquatic Scenes

Based on our aquatic robot, we use DRNet512-VGG16 to detect underwater objects. According to different training domains, we denote detection methods as DRNet512-VGG16-O, DRNet512VGG16-F, and DRNet512-VGG16-G, which are trained on *train, train-F, train-G*, respectively. If DRNet512-VGG16-F or DRNet512-VGG16-G is employed, corresponding visual restoration (i.e., FRS or GAN-RS) should also be adopted to cope with online data. As shown in Figure 6.6, DRNet512VGG16-O almost completely loses its effect on object perception. Besides, DRNet512-VGG16-F and FRS also have difficulty in detecting underwater objects. In contrast, DRNet512-VGG16-G and GAN-RS have higher recall rate and detection precision in this real-world task. Because of the same detection

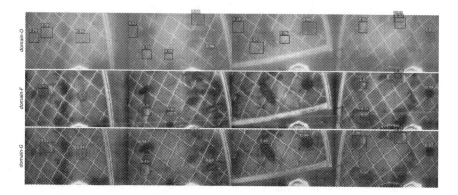

FIGURE 6.6 Demonstration of online detection. DRNet512-VGG16-O and DRNet512-VGG16-F can hardly be qualified for this online detection. By suppressing the problem of domain shift, DRNet512-VGG16-G and GAN-RS perform better in this field underwater scene. Labels of the vertical axis denote training domains. Confidence scores are presented on the top-left of boxes.

method and content of training data, the huge performance gap should be caused by training domain.

6.4.3.2 Online Domain Analysis

As shown in Figure 6.7, there is a huge discrepancy between online domain and *domain-O*. Thus, DRN512-VGG16-*O* suffers from serious degeneration on detection accuracy. Domain shift is moderated by FRS, but FRS

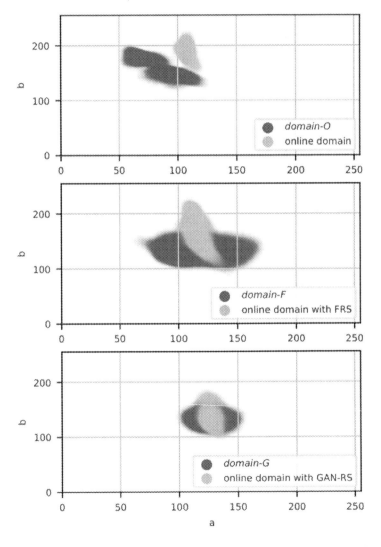

FIGURE 6.7 Comparison of online domain and training domains in Lab color space. Color transparency indicates distribution probability.

is not sufficient to preserve detection performance in this scenario. On the contrary, GAN-RS has higher restoration intensity. As a result, processed by GAN-RS, online domain and *domain-G* are highly overlapped as illustrated in Figure 6.7. Therefore, DRN512-VGG16-G and GAN-RS are able to perform this detection task well. It can be seen that the problem of domain shift is gradually solved with increasing restoration intensity. In addition, underwater scene domains are manifold (see Figure 6.1), so domain-diverse data collection is unattainable. Therefore, contributing to domain shift suppression, visual restoration is essential for object detection in underwater environments.

6.4.4 Discussion

6.4.4.1 Recall Efficiency

In within-domain tests, high-quality domain induces lower detection performance, because of low recall efficiency. Thus, high-quality domain incurs more false positives. However, object candidates that could bring about false positives exist in both training and testing phase. Under this condition, it is seen that the learning of these candidates is insufficient. Therefore, we advocate further research on how these candidates separately impact training and inference for exploring more efficient learning methods.

6.4.4.2 CNN's Domain Selectivity

In cross-domain training, low-quality samples lose their effects so that accuracy drops on *test* set. It is seen that the learning of CNN is characterized by domain selectivity. That is, samples' contributions are different in CNN-based detection learning. Therefore, we advocate further research on CNN's domain selectivity for building more robust real-world detectors.

6.5 UNDERWATER VISION SYSTEM AND MARINE TEST

Based on above analysis and other researches in this book, we develop an underwater robotic system for object perception and grasping tasks. With visual object detection and tracking, the robot can grasp "trepang", "echinus", "shell", etc.

6.5.1 System Design

As shown in Figure 6.8, the aquatic robot is designed for online object detection and grasping. It is 0.68 m in length, 0.57 m in width, 0.39 m in

FIGURE 6.8   Underwater robot for object detection and grasping tasks.

height, and 50 kg in weight. In the robot, we deploy a microcomputer with an Intel I5-6400 CPU, an NVIDIA GTX 1060 GPU, and 8 GB RAM as the processor. In addition, a soft arm is designed to grasp marine animals without damage.

Referring to Figure 6.9, underwater vision system contains three modules:

- Global perception module uses GAN-RS to provide clear vision and fixed data domain.

- Group perception module uses TDRNet [8], OTA [32], and OTR [33] for MOT.

- Target perception module uses TDRNet, SOS [33], and OTR for SOT.

The system is inherited from SOT-by-detection framework [33], and the branch switch conditions are also described in [33].

6.5.2 Underwater Object Counting

As shown in Figure 6.10, we conduct the experiment on underwater object counting. The test venue is Jinshitan, Dalian, China, where the water depth is about 5 m. Based on the group perception module, accurate counting results are obtained.

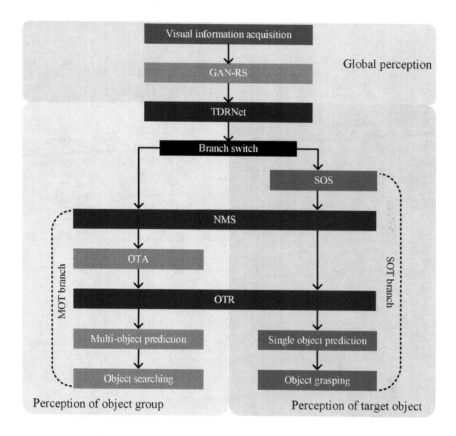

FIGURE 6.9   Underwater vision system for object detection and grasping tasks.

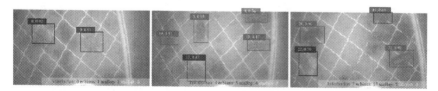

FIGURE 6.10   Underwater object counting. Counting results are shown at the bottom. Objects are described as "ID, confidence score". Object ID is bigger than object counting because OTR discards unstable objects.

6.5.3 Underwater Object Grasping

The proposed vision system is potential and flexible in robotic perception. For example, when a mobile robot tries to grasp all objects in an area, the MOT branch should firstly work for object search, i.e., perception of an object group. After a stable tracklet is detected, the SOT branch should

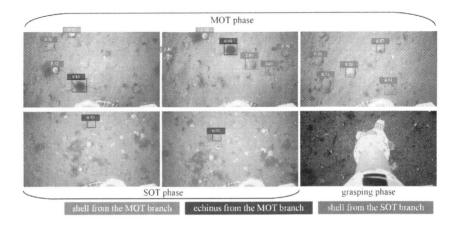

FIGURE 6.11   Underwater object grasping.

work for the elaborate perception of an object instance. At this time, other perception manners (e.g., mask, depth, etc.) can be added for more object instance information to guide this object grasping. If the target encounters tracking failure, our proposed framework can immediately capture this situation and give a convenient way to switch to the MOT branch for object search.

As shown in Figure 6.11, the aquatic robot is developed for grasping marine animals on nature seabed. The test venue is located in Zhangzidao, China, where the water depth is about 15 m. In this task, our vision system is competent in detecting/tracking objects for robotic perception and provides flexible perception ability for object group/target. Based on visual object perception, the robot is able to efficiently approach and grasp targets.

6.6 CONCLUDING REMARKS

In this chapter, we have taken aim at domain analysis based on visual restoration and object detection for underwater robotic perception. Firstly, quality-diverse data domains are derived from URPC2018 dataset with FRS and GAN-RS. Furthermore, single-shot detectors are trained and evaluated, where within-domain and cross-domain performance are unveiled. Finally, we conduct online object detection to reveal the effect of visual restoration on object detection. As a result, we conclude novel viewpoints as follows: (1) Domain quality has an ignorable effect on within-domain convolutional representation and detection accuracy; (2) low-quality domain induces high cross-domain generalization ability; (3) low-quality domain can hardly be well learned in a domain-mixed learning process; (4) visual restoration is a thankless method for elevating within-domain performance, and it incurs relatively low recall efficiency; (5) visual restoration is essential in online robotic perception since it can relieve the problem domain shift. Ultimately, this chapter designs an underwater vision system for object grasping, and underwater object perception and grasping is performed based on underwater robot and soft robotic arm.

In the future, we will further explore domain-related recall efficiency and learning selectivity. Additionally, more robotic tasks will be carried out based on our analysis.

REFERENCES

1. Z. Gong, J. Cheng, X. Chen, W. Sun, X. Fang, K. Hu, Z. Xie, T. Wang, and L. Wen, "A bio-inspired soft robotic arm: Kinematic modeling and hydrodynamic experiments," *J. Bionic Eng.*, vol. 15, no. 2, pp. 204–219, 2018.
2. M. Cai, Y. Wang, S. Wang, R. Wang, Y. Ren, and M. Tan, "Grasping marine products with hybrid-driven underwater vehicle-manipulator system," *IEEE Trans. Autom. Sci. Eng.*, doi: 10.1109/TASE.2019.2957782.
3. J. Gao, A. A. Proctor, Y. Shi, and C. Bradley, "Hierarchical model predictive image-based visual servoing of underwater vehicles with adaptive neural network dynamic control," *IEEE Trans. Cybern.*, vol. 46, no. 10, pp. 2323–2334, 2015.
4. A. Kim and R. M. Eustice, "Real-time visual SLAM for autonomous underwater hull inspection using visual saliency," *IEEE Trans. Rob.*, vol. 29, no. 3, pp. 719–733, 2013.
5. W. Liu, D. Anguelov, D. Erhan, C. Szegedy, S. Reed, C. Y. Fu, and A. C. Berg, "SSD: Single shot multibox detector," in *Proceedings of the European Conference on Computer Vision*, Amsterdam, Netherlands, Oct. 2016, pp. 21–37.
6. S. Zhang, L. Wen, X. Bian, Z. Lei, and S. Z. Li, "Single-shot refinement neural network for object detection," in *Proceedings of the IEEE Conference on Computer Vision and Pattern Recognition*, Salt Lake City, USA, Jun. 2018, pp. 4203–4212.
7. T. Y. Lin, P. Goyal, R. Girshick, K. He, and P. Dollar, "Focal loss for dense object detection," in *Proceedings of the IEEE International Conference on Computer Vision*, Venice, Italy, Oct. 2017, pp. 2980–2988.
8. X. Chen, X. Yang, S. Kong Z. Wu, and J. Yu, "Joint anchor-feature refinement for real-time accurate object detection in images and videos," *IEEE Trans. Circuits Syst. Video Technol.*, doi:10.1109/TCSVT.2020.2980876, 2020.

9. L. Pang, Z. Cao, J. Yu, P. Guan, X. Rong, H. Chai, "A visual leader following approach with a TDR framework for quadruped robots," *IEEE Trans. Syst. Man and Cybern. Syst.*, 2019, doi: 10.1109/TSMC.2019.2912715.

10. Y. Y. Schechner and N. Karpel, "Clear underwater vision," in *Proceedings of the IEEE Conference on Computer Vision and Pattern Recognition*, Washington, USA, Jun. 2004, pages I-536–I-543.

11. R. Liu, X. Fan, M. Zhu, M. Hou, and Z. Luo, "Real-world underwater enhancement: Challenges, benchmarks, and solutions under natural light," *IEEE Trans. Circuits Syst. Video Technol.*, 2020, doi:10.1109/TCSVT.2019.2963772.

12. C. Li, J. Guo, R. Cong, Y. Pang, and B. Wang, "Underwater image enhancement by dehazing with minimum information loss and histogram distribution prior," *IEEE Trans. Image Process.*, vol. 25, no. 12, pp. 5664–5677, 2016.

13. Y. T. Peng and P. C. Cosman, "Underwater image restoration based on image blurriness and light absorption," *IEEE Trans. Image Process.*, vol. 26, no. 4, pp. 1579–1594, 2017.

14. X. Chen, Z. Wu, J. Yu, and L. Wen, "A real-time and unsupervised advancement scheme for underwater machine vision," in *Proceedings of the IEEE International Conference on CYBER Technology in Automation, Control, and Intelligent Systems*, Hawaii, USA, Aug. 2017, pp. 271–276.

15. X. Chen, J. Yu, S. Kong, Z. Wu, X. Fang, and L. Wen, "Towards Real-Time Advancement of Underwater Visual Quality with GAN," *IEEE Trans. Ind. Electron.*, vol. 66, no. 12, pp. 9350–9359, 2019.

16. D.-G. Lowe, "Distinctive image features from scale-invariant keypoints," *Int. J. Comput. Vision*, vol. 60, no. 2, pp. 91–110, 2004.

17. Y. Chen, W. Li, C. Sakaridis, D. Dai, L. Van Gool, "Domain adaptive faster R-CNN for object detection in the wild," in *Proceedings of the IEEE Conference on Computer Vision and Pattern Recognition*, Salt Lake City, USA, Jun. 2018, pp. 3339–3348.

18. J. Xu, S. Ramos, D. Vazquez, and A. M. Lopez, "Domain adaptation of deformable part-based models," *IEEE Trans. Pattern Anal. Mach. Intell.*, vol. 36, no. 12, pp. 2367–2380, 2014.

19. A. Raj, V. P. Namboodiri, and T. Tuytelaars, "Subspace alignment-based domain adaptation for RCNN detector," https://arxiv.org/abs/1507.05578, 2015.

20. M. Khodabandeh, A. Vahdat, M. Ranjbar, and W. G. Macready, "A robust learning approach to domain adaptive object detection," in *Proceedings of the IEEE International Conference on Computer Vision*, Seoul, Korea, Oct. 2019, pp. 480–490.

21. N. Inoue, R. Furuta, T. Yamasaki, and K. Aizawa, "Cross-domain weakly-supervised object detection through progressive domain adaptation," in *Proceedings of the IEEE Conference on Computer Vision and Pattern Recognition*, Salt Lake City, USA, Jun. 2018, pp. 5001–5009.

22. C. Chi, S. Zhang, J. Xing, Z. Lei, S. Z. Li, and X. Zou, "Selective refinement network for high performance face detection," in *Proceedings of the*

AAAI Conference on Artificial Intelligence, Honolulu, USA, Jul. 2019, pp. 8231–8238.

23. Y. Zhu, C. Zhao, H. Guo, J. Wang, X. Zhao, and H. Lu, "Attention couplenet: Fully convolutional attention coupling network for object detection," *IEEE Trans. Image Process.*, vol. 28, no. 1, pp. 113–126, 2019.

24. X. Zhou, J. Zhuo, and P. Krahenbuhl, "Bottom-up object detection by grouping extreme and center points," in *Proceedings of the IEEE Conference Computer Vision and Pattern Recognition*, Long Beach, USA, Jun. 2019, pp. 850–859.

25. V. Kalogeiton, V. Ferrari, and C. Schmid, "Analysing domain shift factors between videos and images for object detection," *IEEE Trans. Pattern Anal. Mach. Intell.*, vol. 38, no. 11, pp. 2327–2334, 2016.

26. M. Yang and A. Sowmya, "An underwater color image quality evaluation metric," *IEEE Trans. Image Process.*, vol. 24, no. 12, pp. 6062–6071, 2015.

27. K. Panetta, C. Gao, and S. Agaian, "Human-visual-system-inspired underwater image quality measures," *IEEE J. Oceanic Eng.*, vol. 41, no. 3, pp. 541–51, 2015.

28. W. H. Lin, J. X. Zhong, S. Liu, T. Li, and G. Li, "RoIMix: Proposal-fusion among multiple images for underwater object detection," *arXiv:1911.03029*, 2019.

29. K. Simonyan and A. Zisserman, "Very deep convolutional networks for large-scale image recognition," *arXiv:1409.1556*, 2014.

30. A. Howard, M. Zhu, B. Chen, D. Kalenichenko, W. Wang, T. Weyand, M. Andreetto, and H. Adam, "Mobilenets: Efficient convolutional neural networks for mobile vision applications," https://arxiv.org/abs/1704.04861, 2017.

31. K. He, X. Zhang, S. Ren, and J. Sun, "Deep residual learning for image recognition," in *Proceedings of the IEEE Conference on Computer Vision and Pattern Recognition*, Las Vegas, USA, Jun. 2016, pp. 770–778.

32. X. Chen, J. Yu, and Z. Wu, "Temporally identity-aware SSD with attentional LSTM," *IEEE Trans. Cybern.*, vol. 50, no. 6, pp. 2674–2686, 2020.

33. X. Chen, J. Yu, Z. Wu, and Li Wen, "Rethinking temporal object detection from robotic perspectives," *arXiv: 1912.10406*, 2019. https://ui.adsabs.harvard.edu/abs/2019arXiv191210406C/abstract

CHAPTER 7

IWSCR

An Intelligent Water Surface Cleaner Robot for Collecting Floating Garbage

7.1 INTRODUCTION

Water is the source of life. The oceans and rivers cover almost 71% of the earth's surface and provide a suitable home for billions of aquatic organisms [1]. However, humanity does not treat the aquatic environment in a friendly manner.

The water pollution resulted from human negligence has been accumulating for decades. The waste in water consists of the dredge, industrial garbage, sewage, radioactive materials, and plastic trash [2]. With expanding applications for robots, many authors have reported their attempts to develop robots designed to clean local environments [3–10]. Naturally, robots designed to clean bodies of water merit development and study.

To collect the plastic pollution, semi-manual refuse-removal vessels are widely applied recently. However, the refuse-removal vessel is large in size, which is only appropriate to rivers with huge acreage or much-accumulated waste. In this context, it is impracticable to clean small, low-density waste in small waters by the refuse-removal vessel. Notably, this method lacks the capacity to determine which floating objects merit removal and which ones do not. Additionally, the exhaust from the refuse-removal vessel may cause secondary pollution. We aim to address these limitations by developing a zero-emissions water-cleaning robot system that can collect objects recognized as garbage.

Above all, the intelligent water surface cleaner robot is a compositive robotic system with vision module, motion control module, and grasping module, which can sequentially accomplish three tasks (TTs), i.e., cruise and detection, tracking and steering, and grasping and collection. In the first task, the robot moves on the water surface following the preplanned path and uses its vision module to detect garbage. When an object is targeted for removal, the second task begins. The vision module tracks the target and measures the relative position between the robot and the target. At the meantime, the motion control module utilizes the position information from the vision module to correct the yaw angle error, seamlessly ensuring an accurate target-approach angle. In the final task, the grasping module determines the grasp timing, and then it commands the manipulator to grasp and collect the objects. Notably, there are three major technical challenges standing in front of accomplishing the aforementioned three tasks:

1. *Challenge for cruise and detection*: To distinguish the garbage despite the robot moving, the object detection algorithm of the cleaner robot should be accurate and real time.

2. *Challenge for tracking and steering*: There are unpredictable dynamic factors in the aquatic environment, like wind and waves, that make it necessary.

3. *Challenge for grasping and collection*: The dynamic conditions on the surface of water means that objects will move, including the robot itself. This poses a challenge for grasping an object that is likely to be moving.

Related theories and methods, which are appropriate to be employed in the cleaner robot to solve the aforementioned problems, have been widely studied in recent years. For the aspect of object detection, deep neural networks have been applied to objects detection in images with great success [11]. There are two major frameworks for detection, i.e., the two-stage framework and the one-stage framework. The two-stage framework is based on R-CNN with a region proposal network [12–15]. With the advantage of high computing speed, the one-stage framework converts the object detection to a bounding box regression problem, such as SSD [16,17] and YOLO [18–20] networks. For techniques of the motion control module, the sliding mode controller (SMC) has been successfully applied on autonomous underwater vehicles (AUVs), owing to its insensitivity to model parameters and external disturbances [21–26]. Additionally, grasping strategies with grasping detection based on the deep neural network are studied extensively nowadays [27–32], by which the graspable position can be determined accurately. However, these data-driven strategies take too much time to execute in real time and are therefore not suitable for the dynamic grasping process.

In this chapter, IWSCR, a prototype of intelligent water surface cleaner robots, is designed, which can work in the experimental tank environment and accomplish the TTs with the purpose of collecting a plastic bottle, a plastic bag, and a Styrofoam floating on the water surface. The IWSCR is a small underwater vehicle that is equipped with a binocular camera, a manipulator, and a collection box. To tackle the challenges, the object detection algorithm based on YOLOv3 network is employed for improving the accuracy and speed of detection by being trained on our proposed dataset; the SMC-based control law with a satisfactory capability of resisting disturbances is designed for the vision-based steering to guide IWSCR toward the target; based on the stability of objects in fluid, a feasible grasping strategy for floating objects is proposed to solve the problem of dynamic grasping. Here, we report our experimental results, which demonstrate that IWSCR has the ability to accomplish TTs and that it is competent to complete the work of water surface cleaner. The primary contributions of this chapter are twofold:

1. IWSCR is a novel design for floating garbage collection. IWSCR accomplishes TTs autonomously due to our integration of computational strategies for vision-based detection, identification, motion control, and grasping. To the best of our knowledge, there are few, if any, published examples of autonomous water surface cleaning robots. We suspect that IWSCR will attract interest from researchers.

2. To overcome the related technical difficulties, the YOLOv3 detection framework, SMC for the vision-based steering, and a feasible grasping strategy based on the stability of objects in a fluid are employed in IWSCR. Experimental results indicate that the performances of

these technical methods are satisfactory. Especially, the proposed grasping strategy inspired by the characteristic of floating objects provides a novel implementation for the dynamic grasping in the fluid. In addition, IWSCR might serve as a platform to test related techniques in the future.

The rest of the chapter is organized as follows. In Section 7.2, the design of IWSCR is overviewed. The object detection framework based on the YOLOv3 network, sliding mode controller for vision-based steering, and grasping strategy are detailed in Sections 7.3, 7.4, and 7.5, respectively. Next, experimental results and discussion are offered in Section 7.6. Finally, the conclusion and future work are summarized in Section 7.7.

7.2 PROTOTYPE DESIGN OF IWSCR

7.2.1 Configuration of IWSCR

The configuration of IWSCR is illustrated in Figure 7.1. The system is based on an electrically powered underwater vehicle that can sail on the surface of the water under load. The underwater vehicle is about 63.5 cm long, 48.5 cm wide, and 46.5 cm high, and it weighs approximately 25 kg. A camera cabin is installed in the front of the top of the vehicle, in which a binocular camera is fixed on the clapboard. A manipulator with 3-DOF is placed on the vehicle, which is composed of three servo motors. Note that the scope of joint angles is 270° to ensure that the tail end of the

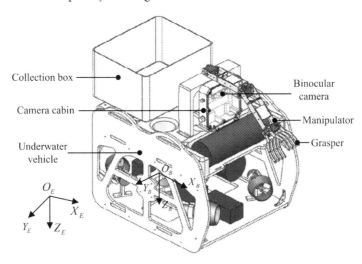

FIGURE 7.1 Configuration of IWSCR.

manipulator can move to the inner of the collection box. Thus, the size of IWSCR is diminutive enough to be applied in the small waters, and its source of energy is clean to avoid secondary pollution.

7.2.2 Framework of Control System

As shown in Figure 7.2, the personal computer (PC) processes with the information from the binocular camera in order to control IWSCR. There are three communication modes between PC and IWSCR, i.e., USB video class (UVC), TCP/IP, and Bluetooth. The image processor in PC obtains images from IWSCR and estimates the objective position. The estimated position flows in two different ways. The first way includes steering error generator, motion controller, and thruster allocation. In this way, the computed error is fed into the motion controller to obtain the relative force and moment. According to the thruster allocation, the force and moment

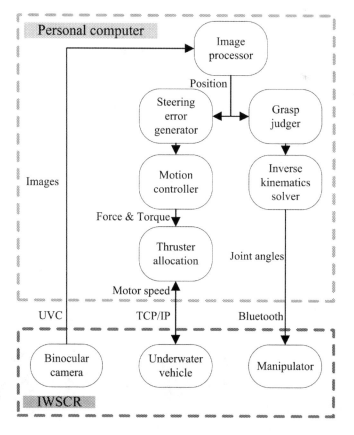

FIGURE 7.2 Framework of control system.

are converted to motor speed, which can control the underwater vehicle directly. The second way includes grasp judger and inverse kinematics solver. In this way, grasping judger determines the timing and graspable position, and the inverse kinematics solver computes the joint angles to control the manipulator. Notice that the control system for IWSCR is inspired by human studies of grasping and object recognition. There is considerable evidence to suggest that humans (and non-human primates) possess different cortical networks for recognizing objects and for planning movements to grasp them [33,34].

7.3 ACCURATE AND REAL-TIME GARBAGE DETECTION

Vision techniques are important for IWSCR to perceive the environment, which provides the basic information for subsequent steering and grasping. Deep learning-based vision techniques perform eminently well in object detection. In this chapter, YOLOv3 network is employed for garbage detection, which is evolved from YOLO and YOLOv2 networks [18–20]. Compared with R-CNN and its ramification networks, YOLO framework transforms the detection problem into a regression problem, which does not require the module of proposal region but generates both of the bounding box coordinates and probabilities of each class directly through regression. Therefore, YOLO framework has greatly higher detection speed compared to Faster R-CNN [18,19], which meets the requirement of real time for IWSCR.

The framework of YOLOv3 based on Darknet-53, shown in Figure 7.3, is applied to detect three classes of objects floating on the water surface, including a plastic bottle, a plastic bag, and a Styrofoam. Different from YOLO and YOLOv2 networks, YOLOv3 uses multi-scale prediction to detect targets and is effective at detecting small objects. For each grid in a different scale, YOLOv3 will predict three bounding boxes. The output in each scale is composed of (1) $\{w, h, x, y, confidence\}$ for each bounding boxes and (2) probabilities of the three classes. Therefore, the total

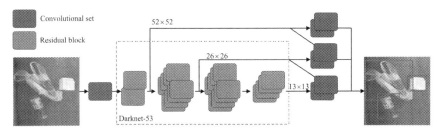

FIGURE 7.3   Framework of YOLOv3 for garbage detection.

channels of output are $3 \times (5+3) = 24$. Among the predicted bounding boxes, the non-maximum suppression (NMS) method is used to select the best bounding box. Besides, to ensure the garbage detection accuracy of the trained network, we have established a floating garbage data set, including almost 1,000 pieces of floating garbage images under different background conditions and illumination intensities. In this context, the total mAP (mean average precision) reaches over than 0.90 in the verification stage, as shown in Section 7.5.

After detection, an object is targeted for removal. Then the kernelized correlation filter (KCF) [35] and the triangulation work to update the bounding box continually and measure the position of the object, respectively.

7.4 SLIDING MODE CONTROLLER FOR VISION-BASED STEERING

In this section, based on the dynamic model of the underwater vehicle, a sliding mode controller is designed for vision-based steering. The proposed control law is robust for disturbance affecting the input.

7.4.1 Dynamic Model of Underwater Vehicle

There is a body-fixed frame $\mathcal{B} = \{X_B, Y_B, Z_B\}$ attached to the vehicle's center of gravity and an inertial frame $\mathcal{I} = \{X_E, Y_E, Z_E\}$ located at the predefined position, as shown in Figure 7.1. Following standard modeling techniques [36–38], the dynamic model of the underwater vehicle in frame $\mathcal{B}$ will be derived according to the general Newton–Euler motion equation in fluid as follows:

$$\mathbf{M}\dot{v} + \mathbf{C}(v)v + \mathbf{D}(v)v + \mathbf{g}(\eta) = \tau_E + \tau$$

$$\dot{\eta} = \mathbf{J}(\eta)v$$

(7.1)

where

- $\eta = \begin{bmatrix} \eta_1 & \eta_2 \end{bmatrix}^T \in \square^6$ denotes the pose vector expressed in $\mathcal{I}$, which involves the position vector $\eta_1 = \begin{bmatrix} x & y & z \end{bmatrix}$ and the orientation vector $\eta_2 = \begin{bmatrix} \phi & \theta & \psi \end{bmatrix}$.

- $v = \begin{bmatrix} v_1 & v_2 \end{bmatrix}^T \in \square^6$ represents the velocity vector expressed in $\mathcal{B}$, which involves the linear velocity vector $[u \ v \ w]$ and the angular velocity vector $v_2 = [p \ q \ r]$.

- $\tau = \begin{bmatrix} \tau_X & \tau_Y & \tau_Z & \tau_K & \tau_M & \tau_N \end{bmatrix}^T \in \square^6$ means the total propulsion vector, i.e., forces τ_X, τ_Y, τ_Z and torques τ_K, τ_M, τ_N generated by thrusters and expressed in frame $\mathcal{B}$.

- $\tau_E \in \square^6$ is the total environmental force/torque vector expressed in frame $\mathcal{B}$, which can be regarded as the external disturbance.

- $M = M_{RB} + M_A$ where $M_{RB} \in \square^{6 \times 6}$ and $M_A \in \square^{6 \times 6}$ are the rigid body and added mass inertia matrices, respectively.

- $C(v) = C_{RB}(v) + C_A(v)$ where $C_{RB}(v) \in \mathbb{R}^{6 \times 6}$ and $C_A(v) \in \square^{6 \times 6}$ are the Coriolis and centripetal matrix which resulted by inertial mass and added mass, respectively.

- $D(v)$ denotes the drag matrix.

- $g(\eta)$ represents the hydrostatic force vector.

- $J(\eta) = diag\{J_1(\eta_2), J_2(\eta_2)\}$ is the Jacobian matrix, transforming velocities from frame $\mathcal{B}$ to frame $\mathcal{I}$, where $J_1(\eta_2)$ denotes the rotation matrix and $J_2(\eta_2)$ stands for lumped transformation matrix.

Considering that the underwater vehicle has three symmetric planes, and that the buoyancy center coincides with the gravity center, the expanded expression of M, $C(v)$, $D(v)$, and $g(\eta)$ are separately described as follows:

$$M = \begin{bmatrix} m - X_{\dot{u}} & 0 & 0 & 0 & 0 & 0 \\ 0 & m - Y_{\dot{v}} & 0 & 0 & 0 & 0 \\ 0 & 0 & m - Z_{\dot{w}} & 0 & 0 & 0 \\ 0 & 0 & 0 & I_x - K_{\dot{p}} & 0 & 0 \\ 0 & 0 & 0 & 0 & I_y - M_{\dot{q}} & 0 \\ 0 & 0 & 0 & 0 & 0 & I_z - N_{\dot{r}} \end{bmatrix}$$

(7.2)

$$\mathbf{C}(v) = \begin{bmatrix} 0 & 0 & 0 \\ 0 & 0 & 0 \\ 0 & 0 & 0 \\ 0 & mw - Z_{\dot{w}}w & -mv + Y_{\dot{v}}v \\ -mw + Z_{\dot{w}}w & 0 & mu - X_{\dot{u}}u \\ mv - Y_{\dot{v}}v & -mu + X_{\dot{u}}u & 0 \end{bmatrix}$$

$$\begin{bmatrix} 0 & mw + Z_{\dot{w}}w & -mv + Y_{\dot{v}}v \\ -mw + Z_{\dot{w}}w & 0 & mu - X_{\dot{u}}u \\ mv - Y_{\dot{v}}v & -mu + X_{\dot{u}}u & 0 \\ 0 & I_z r - N_{\dot{r}}r & -I_y q + M_{\dot{q}}q \\ -I_z r + N_{\dot{r}}r & 0 & I_x p - K_{\dot{p}}p \\ I_y q - M_{\dot{q}}q & -I_x p + K_{\dot{p}}p & 0 \end{bmatrix} \tag{7.3}$$

$$\mathbf{D}(v) = -diag\{X_u - X_{u|u|}|u|, Y_v - Y_{v|v|}|v|,$$

$$Z_w - Z_{w|w|}|w|, K_p - K_{p|p|}|p|, \tag{7.4}$$

$$M_q - M_{q|q|}|q|, N_r - N_{r|r|}|r|\}$$

$$\mathbf{g}(\eta) = \begin{bmatrix} (mg - B)\sin\theta \\ -(mg - B)\cos\theta\sin\phi \\ -(mg - B)\cos\theta\cos\phi \\ 0 \\ 0 \\ 0 \end{bmatrix} \tag{7.5}$$

where

- m, g, and B denote mass, gravity, and buoyancy of the underwater vehicle, respectively.

- $X_{\dot{u}}$, $Y_{\dot{v}}$, $Z_{\dot{w}}$, $K_{\dot{p}}$, $M_{\dot{q}}$, and $N_{\dot{r}}$ are added mass terms.

- I_x, I_y, and I_z are moments of inertia along the related axes.

- X_u, Y_v, Z_w, K_p, M_q, and N_r denote the first-order drag parameters. $X_{u|u|}$, $Y_{v|v|}$, $Z_{w|w|}$, $K_{p|p|}$, $M_{q|q|}$, and $N_{r|r|}$ denote the second-order drag parameters.

In this chapter, the motion of IWSCR is only on the water surface and [ϕ θ] is negligible. Therefore, the motion equation of the underwater vehicle is simplified to three degrees of freedom (i.e., u, v, and r) as follows:

$$\begin{cases} \dot{u}(m - X_{\dot{u}}) - vr(m - Y_{\dot{v}}) - u(X_u + X_{u|u|}\,|u|) \\ \qquad\qquad = \tau_X + \tau_{EX} \\ \dot{v}(m - Y_{\dot{v}}) + ur(m - X_{\dot{u}}) - v(Y_v + Y_{v|v|}\,|v|) \\ \qquad\qquad = \tau_Y + \tau_{EY} \\ \dot{r}(I_z - N_{\dot{r}}) + uv(m - Y_{\dot{v}}) + uv(-m + X_{\dot{u}}) \\ \qquad - r(N_r + N_{r|r|}\,|r|) = \tau_N + \tau_{EN}. \end{cases} \tag{7.6}$$

As for thruster's allocation, each thruster produces forces and torques with respect to frame $\mathcal{B}$ as the following equation:

$$^i\tau = \begin{bmatrix} ^iF \\ ^iQ \end{bmatrix} = \begin{bmatrix} ^ie \\ \left(^iL \times {}^ie\right) \end{bmatrix} {}^iT \tag{7.7}$$

where ^ie denotes the orientation of the ith thruster with respect to frame $\mathcal{B}$, ^iL is the position of attack for ^iT with respect to frame $\mathcal{B}$, and ^iT denotes the thrust of the ith thruster. For the prototype, the distribution of four horizontal thrusters is shown in Figure 7.4, where $^1e = [\cos\beta\ \sin\beta\ 0]^T$, $^2e = [\cos\beta\ -\sin\beta\ 0]^T$, $^3e = [-\cos\beta\ -\sin\beta\ 0]^T$, $^4e = [-\cos\beta\ \sin\beta\ 0]^T$,

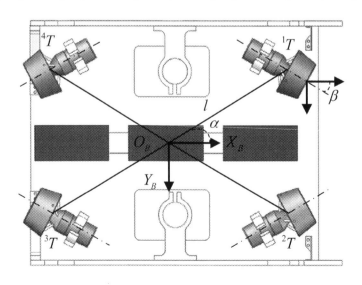

FIGURE 7.4   Distribution of four horizontal thrusters in IWSCR.

$^{1}L = [l\cos\alpha \;-l\sin\alpha \;0]^{T}, ^{2}L = [l\cos\alpha \;l\sin\alpha \;0]^{T}, ^{3}L = [-l\cos\alpha \;l\sin\alpha \;0]^{T}$, and $^{4}L = [-l\cos\alpha \;-l\sin\alpha \;0]^{T}$. By substituting the aforementioned values, the total vector of propulsion forces and torques on the water surface is as follows:

$$\tau_X = \begin{bmatrix} \cos\beta \\ \cos\beta \\ -\cos\beta \\ -\cos\beta \end{bmatrix}^{T} \begin{bmatrix} {}^{1}T \\ {}^{2}T \\ {}^{3}T \\ {}^{4}T \end{bmatrix} \tag{7.8}$$

$$\tau_Y = \begin{bmatrix} \sin\beta \\ -\sin\beta \\ -\sin\beta \\ \sin\beta \end{bmatrix}^{T} \begin{bmatrix} {}^{1}T \\ {}^{2}T \\ {}^{3}T \\ {}^{4}T \end{bmatrix} \tag{7.9}$$

$$\tau_N = \begin{bmatrix} l\cos\alpha\sin\beta + l\cos\beta\sin\alpha \\ -l\cos\alpha\sin\beta - l\cos\beta\sin\alpha \\ l\cos\alpha\sin\beta + l\cos\beta\sin\alpha \\ -l\cos\alpha\sin\beta - l\cos\beta\sin\alpha \end{bmatrix}^{T} \begin{bmatrix} {}^{1}T \\ {}^{2}T \\ {}^{3}T \\ {}^{4}T \end{bmatrix}. \tag{7.10}$$

It is apparent that τ_X, τ_Y, and τ_N can be applied to the underwater vehicle independently with an appropriate combination of thrusts. The rules of thruster's allocation are shown in Table 7.1.

7.4.2 Formulation of the Vision-Based Steering

As shown in Figure 7.5, inspired by [39], the steering objective on the water surface is expressed as follows:

$$\lim_{t \to \infty} \Delta\psi(t) \to 0$$
$$\Delta\psi(t) = \psi_d(t) - \psi(t)$$

where $\psi(t)$ and $\psi_d(t)$ denote the yaw angle of IWSCR and the target yaw angle at time t, respectively. Based on the technique of triangulation for binocular vision, the position of the object $^{C}P(t)$ in the camera-fixed frame

TABLE 7.1   Results of thrusters allocation

Force or Torque	Condition
τ_X^+	$\left({}^1T = {}^2T\right) > \left({}^3T = {}^4T\right)$
τ_X^-	$\left({}^1T = {}^2T\right) < \left({}^3T = {}^4T\right)$
τ_Y^+	$\left({}^1T = {}^4T\right) > \left({}^2T = {}^3T\right)$
τ_Y^-	$\left({}^1T = {}^4T\right) < \left({}^2T = {}^3T\right)$
τ_N^+	$\left({}^1T = {}^3T\right) > \left({}^2T = {}^4T\right)$
τ_N^-	$\left({}^1T = {}^3T\right) < \left({}^2T = {}^4T\right)$

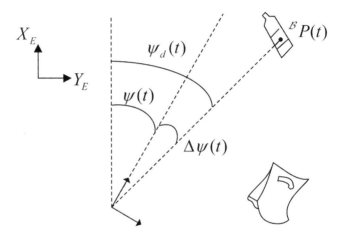

FIGURE 7.5   Formulation of the control objective.

$\mathcal{C}$ can be obtained continually. When ${}^C P(t)$ is transformed to frame $\mathcal{B}$ as ${}^{\mathcal{B}} P(t)$, the difference yaw angle $\Delta \psi(t)$ is derived as follows:

$$\Delta \psi(t) = \arctan \frac{{}^{\mathcal{B}} P(t)_y}{{}^{\mathcal{B}} P(t)_x}. \tag{7.11}$$

The estimated values of velocity u, v, and r are described as follows:

$$\begin{bmatrix} u \\ v \\ r \end{bmatrix} = \begin{bmatrix} -{}^{\mathcal{B}} P(t)_x \\ -{}^{\mathcal{B}} P(t)_y \\ \Delta \psi(t) \end{bmatrix} \tag{7.12}$$

7.4.3 Design and Stability Analysis of Sliding Mode Controller

There are numerous disturbances, which is difficult to identify, against a robot system in an aquatic environment. Besides, the inaccurate measurement of vision system will lead disturbance to the input of the system. Sliding mode controller has the advantages of insensitivity about both of the variance of parameters and disturbance, which is appropriate for IWSCR. Firstly, the sliding surface S is chosen as follows:

$$S = \lambda_0 \Delta \psi(t) + \Delta \dot{\psi}(t) \tag{7.13}$$

where λ_0 is a positive constant. For succinct expression, we define $M_1 = I_z - N_{\dot{r}}$, $M_2 = m - Y_{\dot{v}}$, $M_3 = m - X_{\dot{u}}$, and $M_4 = N_r + N_{r|r|}\,|r|$. From the third equation of (7.6), the control law is constructed as follows:

$$\tau_N = M_1 \left[\frac{(M_2 - M3)}{M_1} uv - \frac{M_4}{M_1} - \lambda_0 r \right] + C_0 sat(S) \tag{7.14}$$

where C_0 is a positive value meeting $C_0 \geq |\tau_{EN}|$, and $sat(S)$ is a saturation function to remove the chattering effect. The expression of saturation function is shown as follows:

$$sat(S) = \begin{cases} 1 & S > \sigma \\ \dfrac{S}{\sigma} & |S| < \sigma \\ -1 & S < -\sigma \end{cases} \tag{7.15}$$

where $\sigma \leq C_0$, a positive value, denotes the boundary layer thickness.

Next, Lyapunov function $V = \dfrac{1}{2} S^2$ is employed to analyze the stability of the designed controller. The derivative Lyapunov candidate function is derived as follows:

$$\dot{V} = S\dot{S}. \tag{7.16}$$

Then, by substituting the expanded expression of $\dot{S}$, $\dot{V}$ can be written as follows:

$$\dot{V} = S[\lambda_0 \Delta \dot{\psi}(t) + \Delta \ddot{\psi}(t)]$$

$$= -\frac{S}{M_1}[M_1 \lambda_0 r - (M_2 - M3)uv \tag{7.17}$$

$$+ M_4 r + (\tau_N + \tau_{EN})].$$

By substituting equation (7.14) in (7.17), $\dot{V}$ can be derived as follows:

$$\dot{V} = \frac{S}{M_1}[-C_0 sat(S) - \tau_{EN}].$$
(7.18)

Consider the saturation function of equation (7.15),

- When $S > \sigma$, $\dot{V} = \dfrac{S}{M_1}[-C_0 - \tau_{EN}] \leq 0$;

- When $S < -\sigma$, $\dot{V} = \dfrac{S}{M_1}[C_0 - \tau_{EN}] \leq 0$;

- When $|S| < \sigma$, we suppose $|\tau_{EN}| \leq \gamma$ and $\dot{V}$ can be derived as follows:

$$
\begin{aligned}
\dot{V} &= -\frac{C_0}{M_1 \sigma} S^2 - \frac{S}{M_1} \tau_{EN} \\
&\leq -\frac{C_0}{M_1 \sigma} S^2 + \frac{|S|}{M_1} \gamma \\
&\leq -\frac{C_0}{M_1 \sigma} S^2 + \frac{1}{2M_1} S^2 + \frac{1}{2M_1} \gamma^2 \\
&= -\left(\frac{C_0}{M_1 \sigma} - \frac{1}{2M_1} \right) S^2 + \frac{1}{2M_1} \gamma^2 \\
&= -\left(\frac{2C_0}{M_1 \sigma} - \frac{1}{M_1} \right) V + \frac{1}{2M_1} \gamma^2 \\
&= -kV + \frac{1}{2M_1} \gamma^2
\end{aligned}
$$
(7.19)

where $k > 0$

Above all, the proposed controller (7.14) is uniformly ultimately bounded (UUB) [40,41]. Therefore, the system is robust for the bounded unknown disturbance.

In addition, the process of engineering realization for vision-based steering is illustrated in Figure 7.6. ψ_T is the threshold value for whether the SMC would be enabled. If $|\Delta\psi(t)| \geq \psi_T$, SMC is applied to generate τ_N in order to reduce $|\Delta\psi(t)|$. If $|\Delta\psi(t)| < \psi_T$, the force τ_X is set to a constant τ_T to urge the underwater vehicle to move forward.

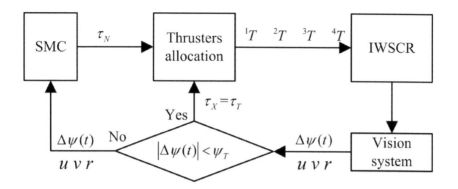

FIGURE 7.6   Process of engineering realization.

7.5 DYNAMIC GRASPING STRATEGY FOR FLOATING BOTTLES

7.5.1 Kinematics and Inverse Kinematics of Manipulator

The manipulator of IWSCR is 3-DOF which can move on a longitudinal vertical plain parallel to the $X_B O_B Z_B$ plain in frame B. The simplified schematic diagram is shown in Figure 7.7. The position of the end effector $^0 P_E(p_x, p_y, 1)$ in the frame $O_0 X_0 Z_0$ can be obtained by

$$^0 P_E = A_1(\mu_1, d_1) \cdot A_2(\mu_2, d_2)^2 P_E \tag{7.20}$$

where

$$A_i(\mu_i, d_i) = \begin{bmatrix} \cos \mu_i & \sin \mu_i & d_i \\ -\sin \mu_i & \cos \mu_i & 0 \\ 0 & 0 & 1 \end{bmatrix},$$

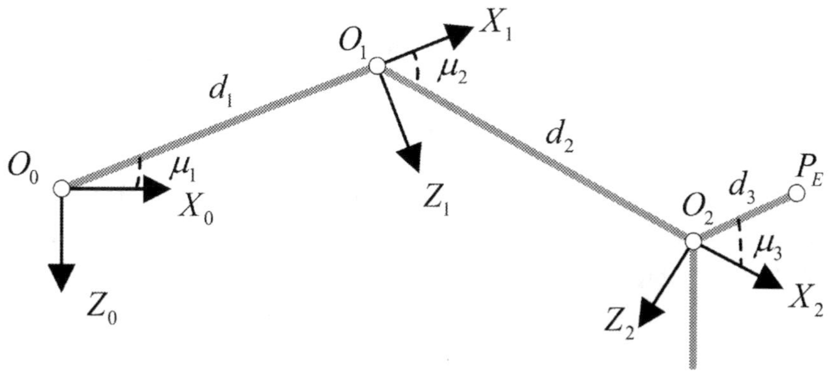

FIGURE 7.7   Simplified schematic diagram of link frames.

$$
^2P_E = \begin{bmatrix} d_3 \cos \mu_3 \\ -d_3 \sin \mu_3 \\ 1 \end{bmatrix}.
$$

In practice, μ_1 and μ_3 are set to constants, so the inverse kinetic equation is as follows:

$$
A_1^{-1}(\mu_1, d_1)^0 P_E = A_2(\mu_2, d_2)^2 P_E. \tag{7.21}
$$

According to the derivation of equation (7.21), μ_2 can be determined by an equation in the following form:

$$
Q_{2\times 2} \begin{bmatrix} \sin \mu_2 \\ \cos \mu_2 \end{bmatrix} = V_{2\times 1}. \tag{7.22}
$$

Additionally, the range of μ_2 is $[-\pi, \pi]$, which ensures the unique solution of equation (7.22). In the process of grasp, when we obtain the position of object, the position and orientation of the manipulator are determined by equation (7.22) simultaneously.

7.5.2 Description of the Feasible Grasping Strategy

In the grasp task, a plastic bag and a Styrofoam are easy to grasp due to their light weight and soft material. Therefore, the bag and the Styrofoam have many graspable positions. However, the plastic bottle is difficult to grasp due to its cylindrical shape, which means the feasible graspable position of plastic bottle locates near the middle of the long axis of the bottle. In order to solve this problem, we proposed a pragmatic grasping strategy for plastic bottles based on the analysis of objects' stability in the fluid. When the vehicle moves toward the bottle, the direction of fluid velocity is from the vehicle to the bottle. In this situation, the bottle will rotate to a stable orientation, which is vertical to the direction of fluid velocity. Therefore, as shown in Figure 7.8, the IWSCR should prepare to grasp in advance, when the distance between IWSCR and the object equals decision distance D_d. Note that D_d is an empirical value related to the present surge speed u and the preparation time T_p, which can be obtained as follows:

$$
D_d = uT_p. \tag{7.23}
$$

Then IWSCR continues moving and grasps the bottle eventually.

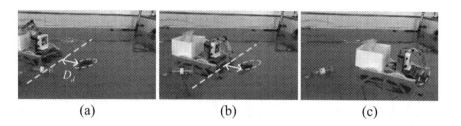

(a) (b) (c)

FIGURE 7.8 Process of the proposed grasping strategy. (a) IWSCR navigates toward the bottle. (b) The manipulator prepares to grasp, when the distance equals D_d. (c) IWSCR continues moving and grasps the bottle.

Next, the proposed strategy will be expounded in theory. Suppose that the plastic bottle could be abstracted as a prolate ellipse on the water surface, that the water velocity v_f is constant, and that the fluid is incompressible. Given the included angle between the long axis of ellipses and the direction of water velocity α_p, refer to [42], the total torque applied on the ellipse is derived as follows:

$$M_E = \frac{1}{2}\pi\rho(a^2 - b^2)v_f^2 \sin 2\alpha_p \qquad (7.24)$$

where ρ denotes the fluid density, a and b are the semi-major axis and the semi-minor axis, respectively. Note that $M_E = 0$, when $\alpha_p = 0$ and $\alpha_p = 1/2\pi$, which represents that the bottle locates at the equilibrium position. However, the stabilities of the two equilibrium states are different.

As to $\alpha_p = 0$, when a disturbance results into an infinitesimal angle $\delta\alpha_p$ adding to the included angle α_p, as shown in Figure 7.9a, the torque applied on the ellipse is as follows:

$$\delta M_E \big|_{\alpha_p = 0} = \pi\rho(a^2 - b^2)v_f^2 \delta\alpha_p \cos 2\alpha_p \big|_{\alpha_p = 0}$$
$$= \pi\rho(a^2 - b^2)v_f^2 \delta\alpha_p \qquad (7.25)$$

where $\delta M_E \cdot \delta\alpha_p > 0$. It means that α_p will increase, so the equilibrium is not stable. Similarly, as to $\alpha_p = 1/2\pi$, shown in Figure 7.9b, the torque applied on the ellipse is as follows:

$$\delta M_E \big|_{\alpha_p = \frac{1}{2}\pi} = \pi\rho(a^2 - b^2)v_f^2 \delta\alpha_p \cos 2\alpha_p \big|_{\alpha_p = \frac{1}{2}\pi}$$
$$= -\pi\rho(a^2 - b^2)v_f^2 \delta\alpha_p \qquad (7.26)$$

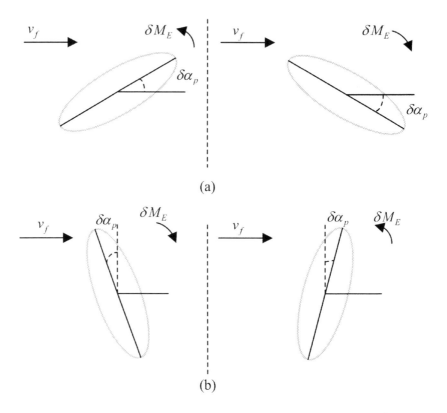

FIGURE 7.9   Stability of the two equilibrium states. (a) $\alpha_p = 0$. (b) $\alpha_p = \dfrac{1}{2}\pi$.

where $\delta M_E \cdot \delta \alpha_p < 0$. It means that α_p will decrease, so the equilibrium is stable.

Above all, the long axis of the plastic bottle will be vertical to the direction of the water velocity, which demonstrates the feasibility of the proposed grasping strategy.

7.6 EXPERIMENTS AND DISCUSSION

In this section, experiments of garbage detection and SMC for vision-based steering are concretely introduced. The whole experimental results of IWSCR to accomplish TTs are exhibited here.

7.6.1 Experimental Results of Garbage Detection

To realize the garbage detection, a floating garbage dataset (FGD) is established, which includes 1,000 images covering various plastic bottles, plastic bags, and Styrofoam under different illumination. FGD is divided into a train dataset (TD) and a verification dataset (VD). The GPU in this experiment is NVDIA-1080. YOLOv3 is trained on TD, and the results on VD are shown in Table 7.2. The accuracy is described with the mean average precision (mAP), and the computing speed is evaluated by frame per second (fps). Note that the detection accuracy is commendable and the speed meets the requirement of real time. However, the detection accuracy of the plastic bottle is a bit lower than the accuracy of the bag and the Styrofoam, for the packaging of plastic bottles are various.

7.6.2 Experimental Results of SMC for Vision-Based Steering and Achievement of TTs

The major advantage of SMC is that it is robust to disturbance caused by the input. Therefore, related simulation experiments based on the underwater vehicle model for the process of steering on the water surface are carried out. In the simulation experiments, the model parameters of IWSCR are listed in Table 7.3. Note that $\Delta\psi(0) = 30°$ is set in the series experiments and that the disturbances τ_{EN} are sine waves with different amplitudes led to the input. The comparison experimental results about the antijamming capability between SMC and PI are shown in Figure 7.10. When none disturbance impacts the system, the proposed controller and PI have good

TABLE 7.2   Results of garbage detection

Class	Accuracy (mAP)	Speed (fps)
Plastic bottle	0.8799	–
Plastic bag	0.9376	–
Styrofoam	0.9182	–
Total	0.9119	45.9036

TABLE 7.3   Model parameters for simulation

Parameter	Value	Parameter	Value
m	25	X_u	−28.5
I_z	20.4	Y_v	−32.5
X_u	−64.3	$X_{u\|u\|}$	−23.6
Y_v	−70.5	$Y_{v\|v\|}$	−42.5
N_r	−55.1	N_r	−10.4
$N_{r\|r\|}$	−18.4	–	–

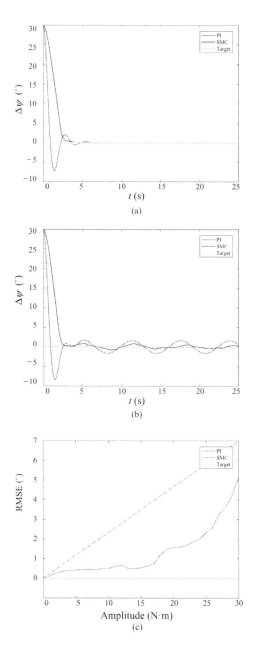

FIGURE 7.10   Comparison experiments between SMC and PI. (a) With none disturbance. (b) Amplitude of disturbance is 5 N m. (c) Relationship between Root Mean Square Error (RMSE) and amplitudes of disturbance.

steady-state performance. However, the PI wastes longer convergence time than the proposed controller. When a sine-type disturbance occurs on the system, the proposed controller is more robust than PI. To describe the antijamming capability of the two controllers more intuitively, we use the Root Mean Square Error (RMSE) to evaluate the steady-state error. As shown in Figure 7.10c, with the disturbance intensities increasing, the RMSEs of the two controllers increase. It is apparent that the RMSE of the proposed controller increases slowly, which demonstrates the SMC's insensibility to disturbance. Additionally, it is difficult for PI controller to avoid overshoot, which is a fatal problem for vision-based steering, because the overshoot may cause the loss of targets in practice. Above all, the results from simulation experiments demonstrate the superiority of SMC applied on IWSCR.

Based on the results of simulation, three experiments of vision-based steering of IWSCR in the real laboratory environment are carried out, as shown in Figure 7.11. The target in each experiment locates by different initial angles relative to IWSCR as shown in Figure 7.11a, and the tendency of the yaw angle error deviation obtained from the binocular vision sensor for the third experiment is described in Figure 7.11b. Notice that the overshoot exists in the experiment, due to the vision measurement errors and the imprecise model parameters. In conclusion, the capability of IWSCR to approach the target demonstrates the feasibility of the proposed controller law.

By applying the detection framework, SMC controller, and the feasible grasping strategy on IWSCR, the performance of autonomously accomplishing TTs is shown in Figure 7.12. For better comprehension, the video is captured by two cameras, i.e., an external camera and IWSCR's binocular camera. For this experiment, the target garbage is a plastic bottle. As shown in Figure 7.12a, IWSCR is cruising on the water surface and detecting garbage by bounding boxes in real time. Due to the objects being plastic bags and Styrofoam, IWSCR does not conduct the tracking and steering task. When the plastic bottle is detected, the target is determined. Figure 7.12b shows that the vision-based steering starts after the target has been locked. Finally, the process of grasp and collection is recorded in Figure 7.12c. The video snapshots demonstrate the feasibility of IWSCR to clean the water surface.

To further evaluate the adaptability of IWSCR, a field experiment of autonomous garbage cleaning has been conducted in a reservoir. The

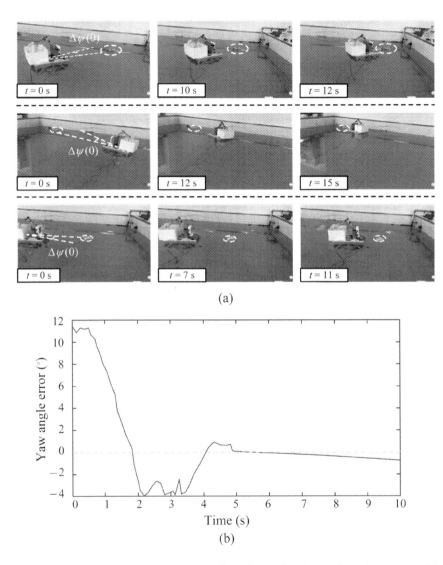

(a)

(b)

FIGURE 7.11 Experiments on vision-based steering in real environment. (a) Video snaps of the vision-based steering with three different initial relative angles. (b) Tendency curve of yaw angle error deviation of the third experiment.

performance of IWSCR is shown in Figure 7.13, where Figure 7.13a records the video snapshots of the cleaning process and Figure 7.13b plots tendency curves for yaw angle error deviation and the distance deviation measured by the binocular vision system. Notice that at $t = 11$, IWSCR got into the grasping preparation stage; thus the visual feedback data was interrupted. An interesting event happened in the cleaning process, that

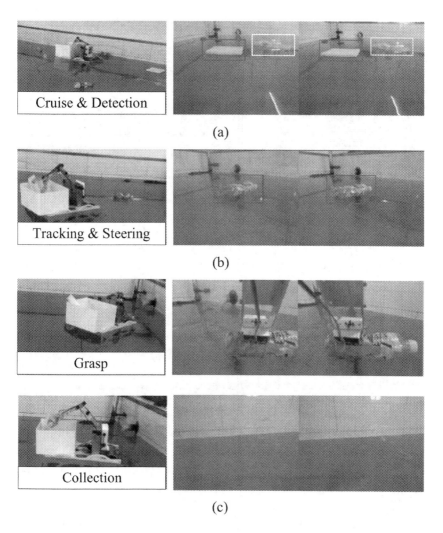

FIGURE 7.12 Performance of accomplishing TTs. (a) Cruise and detection. (b) Tracking and steering. (c) Grasping and collection.

the first grasping attempt failed, but IWSCR detected the target again soon at $t = 18$ and performed the second grasping successfully. Consequently, the field environmental results demonstrate a good utilization prospect of IWSCR for the aquatic environment protection.

7.6.3 Discussion

According to the aforementioned experimental results of IWSCR, our prototype can successfully realize garbage detection, vision-based steering, and water surface grasp in the laboratory environment. During

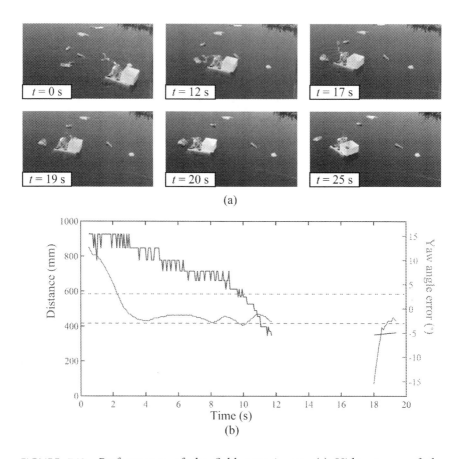

FIGURE 7.13 Performance of the field experiment. (a) Video snaps of the whole cleaning process. (b) Tendency curve of the yaw angle error and distance deviations.

implementation, we overcome three major problems. The first one is how to increase the speed of detection, for which we employ YOLOv3 network to achieve a high accurate and real-time detection whose mAP and speed are up to 0.91 and 45.9 fps, respectively. The second one is how to resist the disturbance in vision-based steering, as to which a control law derived by SMC is proposed. The third one is how to grasp plastic bottles floating on the surface.

The development of an intelligent robot for cleaning water surface is a sophisticated multidisciplinary integration systematic engineering project with numerous challenges. With respect to the visual techniques we employed here for detecting and recognizing objects as garbage, which is the basis of the cleaning, we established specific dataset for garbage floating on the water surface. However, the generality of this dataset is questionable. Furthermore, there are still many aspects to be improved, such as tracking accuracy and match accuracy for triangulation. With respect to controller design, we only tested IWSCR in the laboratory and relatively calm field environments, whereas the expanse water possesses more noise and poses more physical disturbances and other unpredictable influences. Therefore, designing a more robust controller is necessary for achieving success in more complicated real-world settings. With respect to grasping, the proposed strategy is just for specific floating objects. It is necessary to study the dynamic grasp on the water surface further to form a general method.

7.7 CONCLUSION AND FUTURE WORK

In this chapter, IWSCR has been creatively designed to collect plastic garbage on the water surface. To approach the three challenges declared in the introduction (Section 7.1), we employ and propose feasible methodology on IWSCR. Firstly, YOLOv3 network is trained on proposed FGD for garbage detection to ensure the high accuracy and high speed. Secondly, an SMC-based control law is applied to improve the capacity of resisting disturbance for IWSCR. Thirdly, a feasible grasping strategy inspired by the stability of floating object in the fluid is proposed, which provides a novel pathway for a dynamic grasp on the water surface. With aforementioned three techniques, IWSCR possesses the capability to accomplish TTs, and it becomes a pragmatic water surface cleaner.

In the future, IWSCR will be tested in a larger scale and more complicated field environments, which needs much more robust and stable techniques for vision module, motion control module, and grasping module.

REFERENCES

1. J. R. Jambeck, R. Geyer, C. Wilcox, T. R. Siegler, M. Perryman, A. Andrady, R. Narayan, K. L. Law, "Plastic waste inputs from land into the ocean," *Science*, vol. 347, no. 6223, pp. 768–771, 2015.
2. L. C. M. Lebreton, J. Van Der Zwet, J.-W. Damsteeg, B. Slat, A. Andrady, and J. Reisser, "River plastic emissions to the world's oceans," *Nat. Commun.*, vol. 8, p. 15611, 2017.
3. J. Palacin, J. A. Salse, I. Valganon, and X. Clua, "Building a mobile robot for a floor-cleaning operation in domestic environments," *IEEE Trans. Instrum. Meas.*, vol. 53, no. 5, pp. 1418–1424, 2004.

4. M. C. Kang, K. S. Kim, D. K. Noh, J. W. Han, and S. J. Ko, "A robust obstacle detection method for robotic vacuum cleaners," *IEEE Trans. Consum. Electron.*, vol. 60, no. 4, pp. 587–595, 2014.

5. F. Yuan, S. Hu, H. Sun, and L. Wang, "Design of cleaning robot for swimming pools," in *Proceedings of the 2011 International Conference on Management Science and Industrial Engineering (MSIE)*, Harbin, Jan. 2011, pp. 1175–1178.

6. S. Chen, D. Wang, T. Liu, W. Ren, and Y. Zhong, "An autonomous ship for cleaning the garbage floating on a lake," in *Proceedings of the 2009 2nd International Conference on Intelligent Computation Technology and Automation*, Changsha, Oct. 2009, pp. 471–474.

7. H. Zhang, J. Zhang, G. Zong, W. Wang, and R. Liu, "Sky Cleaner 3: A real pneumatic climbing robot for glass-wall cleaning," *IEEE Rob. Autom. Mag.*, vol. 13, no. 1, pp. 32–41, 2006.

8. G. Ferri, A. Manzi, P. Salvini, B. Mazzolai, C. Laschi, and P. Dario, "DustCart, an autonomous robot for door-to-door garbage collection: From DustBot project to the experimentation in the small town of Peccioli," in *Proceedings of the 2011 IEEE International Conference on Robotics and Automation*, Shanghai, May 2011, pp. 655–660.

9. S. Watanasophon and S. Ouitrakul, "Garbage collection robot on the beach using wireless communications," in *Proceedings of the 2014 3rd International Conference on Informatics, Environment, Energy and Applications*, Singapore, Mar. 2014, pp. 92–96.

10. J. Bai, S. Lian, Z. Liu, K. Wang, and D. Liu, "Deep learning based robot for automatically picking up garbage on the grass," *IEEE Trans. Consum. Electron.*, vol. 64, no, 3, pp. 382–389, 2018.

11. L., Yann, Y. Bengio, and G. Hinton, "Deep learning," *Nature*, vol. 521, no. 7553, pp. 436–444, 2015.

12. R. Girshick, J. Donahue, T. Darrell, and J. Malik, "Rich feature hierarchies for accurate object detection and semantic segmentation," in *Proceedings of the IEEE Conference on Computer Vision and Pattern Recognition*, Columbus, USA, Jun. 2014, pp. 580–587.

13. R. Girshick. "Fast R-CNN," in *Proceedings of the IEEE International Conference on Computer Vision*, Santiago, Chile, Dec. 2015, pp. 1440–1448.

14. S. Ren, K. He, R. Girshick, and J. Sun, "Faster R-CNN: Towards realtime object detection with region proposal networks," in *Proceedings of the Advances in Neural Information Processing Systems*, Montreal, Canada, Dec. 2015, pp. 91–99.

15. K. He, G. Gkioxari, P. Dollar, and R. Girshick, "Mask R-CNN," in *Proceedings of the International Conference on Computer Vision*, Venice, Italy, Oct. 2017, pp. 2961–2969.

16. W. Liu, D. Anguelov, D. Erhan, C. Szegedy, S. Reed, C. Y. Fu, and A. C. Berg, "SSD: Single shot multibox detector," in *Proceedings of the European Conference on Computer Vision*, Amsterdam, Netherlands, Oct. 2016, pp. 21–37.

17. X. Chen, Z. Wu, and J. Yu, "TSSD: Temporal single-shot detector based on attention and LSTM," in *Proceedings of the IEEE/RSJ International Conference on Intelligent Robots and Systems*, Madrid, Spain, Oct. 2018, pp. 5758–5763.

18. J. Redmon, S. Divvala, R. Girshick, and A. Farhadi, "You only look once: Unified, real-time object detection," in *Proceedings of the IEEE Conference on Computer Vision and Pattern Recognition*, Las Vegas, USA, Jun. 2016, pp. 779–788.

19. J. Redmon and A. Farhadi, "YOLO9000: Better, faster, stronger," in *Proceedings of the IEEE Conference on Computer Vision and Pattern Recognition*, Honolulu, USA, Nov. 2017, pp. 6517–6525.

20. J. Redmon and A. Farhadi, "YOLOv3: An incremental improvement," https://arxiv.org/abs/1804.02767, 2018.

21. R. Cui, X. Zhang, and D. Cui, "Adaptive sliding-mode attitude control for autonomous underwater vehicles with input nonlinearities," *Ocean Eng.*, vol. 123, pp. 45–54, 2016.

22. Z. Yan, H. Ju, and S. Hou, "Diving control of underactuated unmanned undersea vehicle using integral-fast terminal sliding mode control," *J. Cent. South Univ.*, vol. 23, no. 5, pp. 1085–1094, 2016.

23. H. Zhou, K. Liu, Y. Li, and S. Ren, "Dynamic sliding mode control based on multi-model switching laws for the depth control of an autonomous underwater vehicle," *Int. J. Adv. Robot. Syst.*, vol. 12, no. 7, pp. 1–10, 2015.

24. R. Cui, L. Chen, C. Yang, and M. Chen, "Extended state observer-based integral sliding mode control for an underwater robot with unknown disturbances and uncertain nonlinearities," *IEEE Trans. Ind. Electron.*, vol. 64, no. 8, pp. 6785–6795, 2017.

25. J. Yu, J. Liu, Z. Wu, and H. Fang, "Depth control of a bioinspired robotic dolphin based on sliding-mode fuzzy control method," *IEEE Trans. Ind. Electron.*, vol. 64, no. 3, pp. 2429–2438, 2018.

26. G. P. Incremona, G. D. Felici, A. Ferrara, and E. Bassi, "A supervisory sliding mode control approach for cooperative robotic system of systems," *IEEE Syst. J.*, vol. 6, no. 1, pp. 263–272, 2015.

27. Lenz, H. Lee, and A. Saxena, "Deep learning for detecting robotic grasps," *Int. J. Robot. Res.*, vol. 34, no. 4–5, pp. 705–724, 2015.

28. S. Kumra and C. Kannan, "Robotic grasp detection using deep convolutional neural networks," in *Proceedings of the IEEE/RSJ International Conference on Intelligent Robots and Systems*, Vancouver, Canada, Sep. 2017, pp. 769–776.

29. J. Redmon and A. Angelove, "Real-time grasp detection using convolutional neural networks," in *Proceedings of the International Conference on Robotics and Automation*, Seattle, Washington, USA, May 2015, pp. 1316–1322.

30. U. Asif, M. Bennamoun, and F. A. Sohel, "RGB-D object recognition and grasp detection using hierarchical cascaded forests," *IEEE Trans. Rob.*, vol. 33, no. 3, pp. 547–564, Jun. 2017.

31. E. Johns, S. Leutenegger, and A. J. Davison, "Deep learning a grasp function for grasping under gripper pose uncertainty," in *Proceedings of the IEEE/*

RSJ International Conference on Intelligent Robots and Systems, Daejeon, Korea, Oct. 2016, pp. 4461–4468.

32. U. Asif, J. Tang, and S. Harrer, "GraspNet: an efficient convolutional neural network for real-time grasp detection for low-powered devices," in *Proceedings of the International Joint Conference on Artificial Intelligence*, Stockholm, Sweden, Jul. 2018, pp. 4875–4882.

33. M. A. Goodale, J. P. Meenan, H. H. Bulthoff, D. A. Nicolle, K. J. Murphy, and C. I. Racicot, "Separate neural pathways for the visual analysis of object shape in perception and prehension," *Curr. Biol.*, vol. 4, no. 7, pp. 604–610, 1994.

34. R. L. Whitwell, A. D. Milner, and M. A. Goodale, "The two visual systems hypothesis: new challenges and insights from visual form agnosic patient DF," *Front. Neurol.*, vol. 5, pp. 255, 2014.

35. J. F. Henriques, R. Caseiro, P. Martins, and J. Batista, "High-speed tracking with kernelized correlation filters," *IEEE Trans. Pattern Anal. Mach. Intell.*, vol. 37, no. 3, pp. 583–596, 2015.

36. T. Fossen, *Guidance and Control of Ocean Vehicles*. New York, NY, USA: Wiley, 1994.

37. G. C. Karras, P. Marantos, C. P. Bechlioulis, and K. J. Kyriakopoulos, "Unsupervised online system identification for underwater robotic vehicles," *IEEE J. Oceanic Eng.*, vol. 99, pp. 1–22, 2018, doi: 10.1109/JOE.2018.2827678.

38. Z. Peng and J. Wang, "Output-feedback path-following control of autonomous underwater vehicles based on an extended state observer and projection neural networks," *IEEE Trans. Syst. Man Cybern. Syst.*, vol. 48, no. 4, pp. 535–544, 2018.

39. F. Ke, Z. Li, H. Xiao, and X. Zhang, "Visual servoing of constrained mobile robots based on model predictive control," *IEEE Trans. Syst. Man Cybern. Syst.*, vol. 47, no. 7, pp. 1428–1438, 2017.

40. J. Sun, Z. Pu, and J. Yi, "Conditional disturbance negation based active disturbance rejection control for hypersonic vehicles," *Control Eng. Pract.*, vol. 84, pp. 159–171, 2019.

41. J. Sun, J. Yi, Z. Pu, and X. Tan, "Fixed-time sliding mode disturbance observer-based nonsmooth backstepping control for hypersonic vehicles," *IEEE Trans. Syst. Man Cybern. Syst.*, vol. 50, no. 11, pp. 4377–4386, Nov. 2020. DOI: 10.1109/TSMC.2018.2847706.

42. L. N. Milne-Thomson, *Theoretical Hydrodynamics*. North Chelmsford, MA: Courier Corporation, 1996.

CHAPTER 8

Underwater Target Tracking Control of an Untethered Robotic Fish with a Camera Stabilizer

8.1 INTRODUCTION

Recent years have witnessed an increase in interests in the development and deployment of bioinspired aquatic mechatronic systems [1–4]. Motivated by the prominent swimming abilities of fish, a multitude of efforts have been devoted to the development of fishlike robots, termed robotic fish. On the one hand, fish have evolved amazing adaptations to their environments, offering a range of design options in highly dynamic aquatic environments. On the other hand, integrating biological features into aquatic robotic systems creates favorable opportunities for enhanced understanding of fish propulsion and maneuvering [5]. In particular, maneuverability, efficiency, and stealth performance are three key factors that differentiate bioinspired robotic fish from other types of aquatic robots [6–9]. Undoubtedly, robotic fishes hold tremendous promise for real-world applications, such as oceanography, surveillance, archaeology, patrol, marine environmental monitoring, and mobile sensing, where operations are highly dangerous or impractical for humans or conventional underwater vehicles.

At present, the vast majority of existing studies of robotic fish involve propulsive mechanisms of fish swimming, actuations, mechanical structures, motion control, cooperative control, and so forth [7, 8, 10–12]. Results on vision-guided motion control for untethered robotic fish, however, are still limited owing to the difficulties in high-quality underwater imaging and accurate dynamic modeling of fish swimming. For example, Hu et al. performed vision-based target tracking and collision avoidance in 2D space for a boxfish-like swimming robot using median-paired fin (MPF) swimming [13]; Takada et al. estimated the vision-based target location performance for a single-joint robotic fish [14]; Yu et al. investigated 3D tracking control of a body/caudal fin (BCF)–type robotic fish with embedded vision [15]. Despite being a dominating sensor for monitoring underwater scenes, there are still several unsolved issues with existing underwater cameras and processing methods. Because of nonuniform illumination and unique light transport characteristics in water, the acquired underwater images often suffer from scatters and a large amount of noise. Another constraint for an untethered and self-powered robotic fish is that the computational capacity of its embedded system is limited. Furthermore, intrinsic oscillatory motions in fishlike propulsion inevitably cause sway of the onboard camera mounted in the foremost part of the fish head, which is accompanied by severe degeneration of underwater images. Note that the BCF-type robotic fish is generally inferior to the MPF-type one in maneuverability and stability [5]. Hence, the image degeneration of the BCF-type robotic fish is worse than that of the MPF-type robotic fish.

In order to tackle the problem of the instability of image, one commonly used approach is to alleviate the passive sway effect on the camera. Sun et al. attempted to reduce the sway of the robotic fish head by optimizing the parameters of swimming motions in the BCF mode [16]. In their method, the instability of image was decreased at the sacrifice of swimming speed. Another alternative is to utilize a pan-and-tilt camera. Using an unmanned aircraft system as the verification platform, Doyle et al. presented an algorithm based on optical flow to track a moving object with a real-time pan/tilt camera [17]. Similarly, the concept of a pan-and-tilt camera can be transferred to stabilize the underwater images. Certainly, the realization of the pan-and-tilt camera in a robotic fish is subject to a fusiform body shape and limited computational resource within the bio-inspired design framework.

Considering that the interaction between the fish body and the surrounding fluid which is nonlinear and time-varying, it is very difficult to establish an accurate dynamic model. Further, parametric uncertainties and external disturbances in dynamic aquatic environments cause problems in deriving control laws based on an inaccurate dynamic model. Meanwhile, reinforcement learning (RL), as a method of machine learning, which learns the strategy that best suits the current environment [18], is an appropriate option for the motion control of robotic fish. RL can estimate the unknown environments. In 1968, Michie et al. initially used the Monte Carlo method to estimate the action-value function [19]. Since then, theory and applications of RL are extensively investigated. Lillcrap et al. proposed a deep RL algorithm, called deep deterministic policy gradient (DDPG), which can generate strategies for continuous action space [20]. The utilization of RL in the motion control of robotic fish has two notable advantages: (1) The use of model-independent RL methods can reduce the need for complex hydrodynamic modeling of robotic fish; and 2) Thanks to the self-evolution characteristics of RL, robotic fish can gradually adapt to environmental changes. For instance, Shen et al. used an RL algorithm to enhance the autonomy of a bionic robotic fish [21]. Lin et al. employed a supervised neural Q-learning algorithm to achieve the stability control and trajectory tracking of bionic underwater robots [22]. In practice, RL encounters many difficulties when applied to robotic systems, such as dimension failure, insufficient sample, and return function selection. Even so, the use of RL for robot control is gaining great popularity in a wide spectrum of robotic applications over the last decade.

The main purpose of this chapter is to develop an integrated and feasible solution to underwater target tracking using an untethered vision-guided robotic fish. To this end, we firstly develop a novel active visual tracking system with the inclusion of a camera stabilizer, which can stably locate the tracked target in the camera's field of view (FoV). Then, a DDPG-based target tracking controller is proposed, allowing the continuous tracking of a selected target. In particular, the state vector, the action vector, and the return function are designed reasonably by the analysis of the controlled system and the traced target. The effectiveness of the proposed RL-based target tracking is demonstrated in simulation and experimental results. Compared with the previous research on the tracking control of robotic fish

[13–16], a camera stabilizer with a feedback-feedforward controller is first created, allowing stable and clear underwater image acquisition. Besides, the presented RL-based tracking parameter tuning approach largely reduces the dependency on accurate dynamic models of fish swimming. According to the authors' knowledge, this is the first time that wide-range moving target tracking is achieved on the self-propelled BCF-type robotic fish.

This chapter introduces an excellent project about how to coordinate underwater vision technologies with underwater vehicle control methods. The proposed vision-based control framework applies KCF, one of the state-of-the-art visual tracking methods, the pan-tilt-based image stabilization technique, and the mainstream deep learning concept. In the context, this chapter can inspire readers to study underwater vision applications further.

The remainder of this chapter is organized as follows. The overall mechatronic design and motion control of the developed robotic fish prototype are provided in Section 8.2. The active vision tracking system is described in Section 8.3. Section 8.4 details the RL-based target tracking control method. Section 8.5 offers the experimental setup and corresponding results. Finally, concluding remarks are given in Section 8.6.

8.2 SYSTEM DESIGN OF THE ROBOTIC FISH WITH A CAMERA STABILIZER

This section presents a newly developed BCF-type robotic fish with a camera stabilizing system for stable underwater image acquisition. Using a bioinspired central pattern generator (CPG) as the system actuation, the robotic fish is able to produce multimodal smooth swimming motions.

8.2.1 Mechatronic Design

Inspired by the natural shark, a fast-swimming ocean predator, a shark-like robotic fish is developed [23]. As illustrated in Figure 8.1(a), the robotic fish is comprised of a rigid anterior body with one pair of pectoral fins and a multi-link posterior body with a heterocercal caudal fin. To obtain a better underwater perceptual ability, an embedded vision system with a camera stabilizer is built. For a BCF-type swimmer in motion, the fish head always sways inevitably due to the recoil effect from the left-to-right body undulation, which may lead to severe image blurring. Furthermore, the camera might lose the tracked target during the swaying because of the small field of vision. Aiming at reducing image blur and obtaining

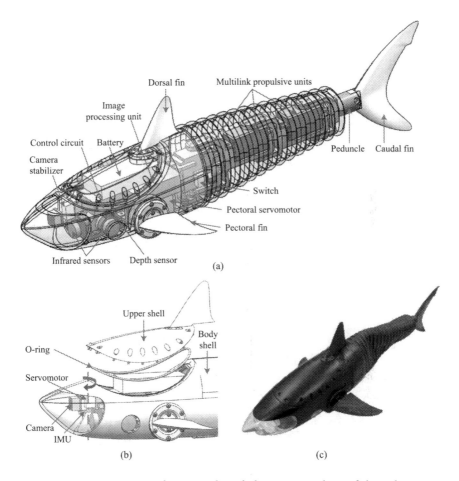

FIGURE 8.1   Illustration of an untethered three-joint robotic fish with a camera stabilizer. (a) Mechanical diagram. (b) Feature of the camera stabilizer. (c) Photograph of the robotic prototype.

a larger field of vision, we propose a novel mechanism for stabilizing the camera, i.e., a camera stabilizer. As depicted in Figure 8.1(b), the camera stabilizer has a motor rotating around the yaw axis in the range of −70° to 70°, which is responsible for modulating the yaw angle of the camera. In general, a pan-and-tilt camera has at least two DoFs. However, owing to the strict interior space limitation on the robot head, a two-DoF camera stabilizer is unavailable. By comparison, the sway around the yaw axis is more violent than other axes. Finally, a scheme of a one-DoF camera stabilizer is eventually adopted, consisting of an attitude measurement unit and a servomotor rotating around the yaw axis. Specifically, two inertial

TABLE 8.1 Technical specification of the developed robotic prototype

Items	Characteristics
Size ($L \times W \times H$)	Approx. 483 mm × 208 mm × 125 mm
Total mass	Approx. 1.3 kg
Joint configuration	Posterior body: 3; Pectoral fins: 2
Onboard sensor	Infrared detectors, pressure sensor, IMU, and camera
Camera resolution	640 × 480 pixels
Drive mode	Servomotors (HS-7940TH, SAVOX-1251TG)
Controller	ARM Cortex-M4
Control mode	Wireless (RF, 433~MHz) or autonomous mode
Power supply	7.4 V rechargeable Li-Polymer batteries

measurement units (IMUs) are separately mounted on the camera and the fish body to obtain the yaw angles of the camera and the robot body, which can provide feedback for controlling the camera stabilizer. The resulting robotic prototype is shown in Figure 8.1(c). The detailed technical specifications of the robotic fish are tabulated in Table 8.1. It is worth noting that the tracking control implemented in this chapter is only in 2D plane. Thus, there is no necessity to control the pectoral fins for 3D maneuvers in the context of planar tracking control.

8.2.2 CPG-Based Motion Control

In this chapter, a CPG-based motion control method is employed to govern the swimming motions of the robotic fish. The interested reader is referred to [24, 25] for an extensive understanding of the CPG-based control method. For simplicity, we directly provide the applied CPG model of a robotic shark. As shown in Figure 8.2, the CPG model in this chapter is comprised of two chains of amplitude-controlled phase oscillators with controllable nearest-neighbor coupling, where the couple of left and right nodes decides the output angle of the corresponding joint together. The dynamics of the ith ($i = 1, 2, \ldots, 6$) oscillator is described by the following nonlinear differential equation [26].

$$
\begin{cases}
\dot{\theta}_i = 2\pi v_i + \sum_{j \in T(i)} \omega_{ij} \sin\left(\theta_j - \theta_i - \phi_{ij}\right) \\
\ddot{\zeta}_i = a_i \left(\frac{a_i}{4} \left(R_i - \zeta_i \right) - \dot{\zeta}_i \right) \\
x_i = \zeta_i \left(1 + \cos\theta_i \right)
\end{cases}
\tag{8.1}
$$

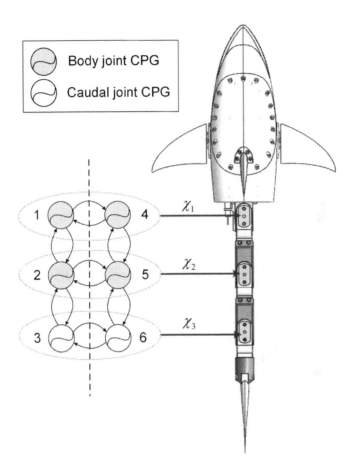

FIGURE 8.2   CPG model applied to the robotic fish.

where θ_i and ζ_i denote the phase and amplitude of the ith oscillator; v_i and R_i represent the intrinsic frequency and amplitude; a_i is a positive constant; ω_{ij} and ϕ_{ij} determine the weights and phase biases of coupling between the ith and jth oscillators. $T(i)$ is the set of oscillators that the ith oscillators receive inbound couplings form. x_i is the rhythmic output of the ith oscillator. The output angle χ_i of the ith joint can be calculated as

$$\chi_i = x_i - x_{3+i}. \tag{8.2}$$

Based on the settings in [23], the CPG model is able to yield a signal for the set point of the angle of servomotor of each joint, which asymptotically converges to the following equation:

$$\chi_i^\infty(t) = (A_{Li} - A_{Ri}) + (A_{Li} + A_{Ri})\cos(2\pi v t + i\Delta\phi + \phi_0) \tag{8.3}$$

where A_{Li} and A_{Ri} represent the intrinsic amplitudes of the left and right oscillators, $\Delta\phi$ indicates the phase bias between oscillators for the descending connections, and ϕ_0 depends on the initial states of the oscillators.

A careful inspection of equation (8.3) reveals that $A_{Li} - A_{Ri}$ stands for the offset while $A_{Li} + A_{Ri}$ for the amplitude of the ith joint. Defining the offset and amplitude as β_i and A_i, it follows that $A_{Li} = \left(A_i + \beta_i\right)/2$ and $A_{Ri} = \left(A_i - \beta_i\right)/2$. Generally, the offset parameter affects the output signals' amplitude offsets, which decide the joints of the robotic fish oscillating around a straight axis ($\beta_i = 0$) or an arc axis ($\beta_i \neq 0$). Obviously, a straight-axis oscillation leads to forward swimming and an arc-axis oscillation leads to turning. Therefore, we can realize the yaw control of the robotic fish by adjusting the offset parameter β_i.

As a consequence, we can adjust the frequency, phase lag, amplitude, and offset of the sinusoidal signal for generating diversified swimming motions by changing v, $\Delta\phi$, A_i, and β_i. Unless otherwise specified, we set $\omega = 4$, $a = 100$, $A_1 = 20\pi/180$, $A_2 = 29\pi/180$, $A_3 = 38\pi/180$ in the subsequent simulations and experiments.

8.3 ACTIVE VISION TRACKING SYSTEM

Target tracking is one of the important tasks for autonomous robots to maintain aim at a static target or monitor the motions of a moving target. In order to accomplish the vision-based target tracking task, besides sufficient locomotion ability, the robotic fish should obtain the information on the target position. It implies that the robotic fish also needs to have the ability to visually track the target. In this work, a small camera is employed to capture underwater images and kernelized correlation filter (KCF) algorithm [27], which has remarkable performance while low computation cost is integrated with different controllers to effectively track a target object. Hence two control efforts are made to construct a stable active vision tracking system.

On the one hand, as shown in Figure 8.3a, a feedback-feedforward controller for the camera stabilizer is built to guarantee image stability. The main concept is to make the camera stabilizer actively rotate to the direction of the target to compensate for the adverse recoil effect on the fish head. It is equivalent to keep the attitude angle of the camera unchanged with respect to the inertial frame as much as possible so that the target will be at the center of the image no matter how the robotic fish moves. For the

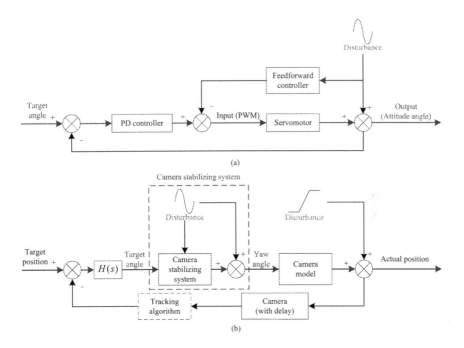

FIGURE 8.3 Control framework of the active vision tracking system. (a) Block diagram of feedback-feedforward control for a camera stabilizing system. (b) Block diagram of image-based tracking control system with the camera stabilizer.

controller design, the main task is to tackle the disturbance on the output, which is induced by the oscillatory fish body. In particular, the disturbance is assumed to be sinusoidal, which can be measured by the onboard IMU. Thanks to the effectiveness of reducing the influences of disturbances, feedforward control is combined with the feedback control being implemented with a proportional-derivative (PD) controller. It should be remarked that the PD controller rather than a proportional-integral-derivative (PID) controller is employed for a better dynamic performance.

On the other hand, to deal with the time-delay problem of embedded vision, as illustrated in Figure 8.3b, a cascade control system is developed. The onboard underwater image acquisition and processing often encounter a severe delay. Contributing factors for this delay include the hardware-related processing time and the implementation time for visual tracking algorithm [28]. When left untreated, the delay might be approximately 1 s. However, the occurred image jitter at a frequency of over 1 Hz is not tolerable. Using the feedback-feedforward controller as an inner loop and the active tracking controller as an outer loop, we rely on cascade control to mitigate the impact of image jitter.

The performance of the camera stabilizing system is demonstrated in Figure 8.4. When there was a constant disturbance represented by a sinusoidal wave with a frequency of 1 Hz and an amplitude of $\pi/6$, the output of the camera stabilizing system was approximately 6°, equivalent to 20% of the disturbance amplitude. Once the camera stabilizing system was required to track a target angle represented by a step signal with an

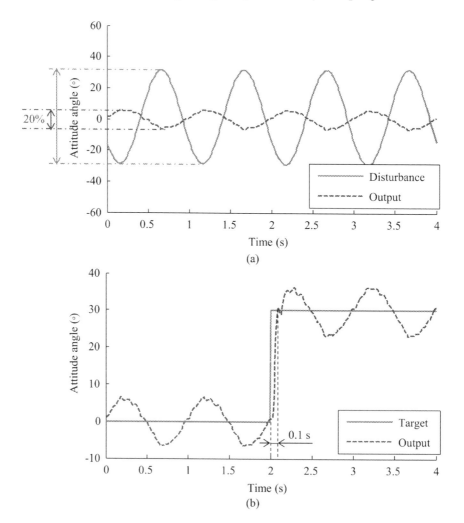

FIGURE 8.4 Response properties of the camera stabilizing system with disturbance. (a) Stable response. (b) Step response.

amplitude of π/6, it took about 0.1 s to arrive at the target attitude angle. If the time to follow the output waveform was considered, the response time of the camera stabilizing system became longer, around 0.2 s, which was acceptable in real visual tracking applications.

In another test, we compared the tracking errors of the active tracking system in simulations and experiments. As shown in Figure 8.5a, the ordinate represents the position of the target object in the image. From $t=5$s, a ramp signal with a duration of 2 s and a slope of 200 pixels per second was exerted on the system as a disturbance. The active visual tracking system controller was able to ensure that the absolute value of the error $|e_{ss}| < 200$, and ultimately there

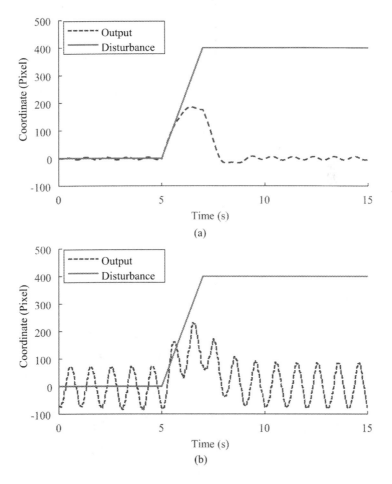

FIGURE 8.5 Tracking errors of active tracking system in simulation and experiment. (a) Simulation result. (b) Experimental result.

was no static error in simulations. As illustrated in Figure 8.5b, the maximum ordinate error was 230 pixels in experiments. It took about 4 s to resume a stable sinusoidal wave, agreeing well with the simulated time. Remarkably, the target was correctly detected and always in the camera's FoV.

Furthermore, for the sake of verifying the performance of the active vision tracking system in an underwater environment, two snapshot cases captured by the underwater camera of robotic fish are compared. Figure 8.6a corresponds to the case without active vision tracking system, whereas Figure 8.6b is the case with active vision tracking system. As can be easily observed, the case with active vision tracking system, in which the target is primarily maintained at the center of snapshot sequence, demonstrates better stability than the case without active vision tracking system. Besides, the images in Figure 8.6a tend to be degraded by motion blur, largely causing the failure of the target detection and tracking. This reveals that the active tracking system was able to reduce the instability of the image and to track the target object even with a large delay during underwater image acquisition and processing, laying a solid foundation for autonomous target tracking control.

8.4 RL-BASED TARGET TRACKING CONTROL

On the basis of the well-developed motion control and active vision tracking system, it is possible to accomplish the target tracking task for robotic fish aided by RL. In this work, a deep Q-learning algorithm that is well suited to continuous control, called DDPG, is exploited as the tracking control algorithm. DDPG is an algorithm which uses off-policy data and the Bellman equation to learn the Q-function and utilizes the Q-function to learn the policy. It is based on the Actor-Critic structure. Specifically, the DDPG algorithm uses two neural networks to replace the Actor and Critic function in deterministic policy gradient (DPG). Actor network takes the current state as input and the action as output. By taking the state and action as input, the output of Critic network is the score of Action network performance and is used to update the weights of Action network. The DDPG algorithm flow chart is illustrated in Algorithm 1.

8.4.1 Tracking Control Design

With the aid of the active vision tracking system, the camera of the robotic fish can be automatically turned toward the target object through the camera stabilizer. As illustrated in Figure 8.7, the relative position relationship

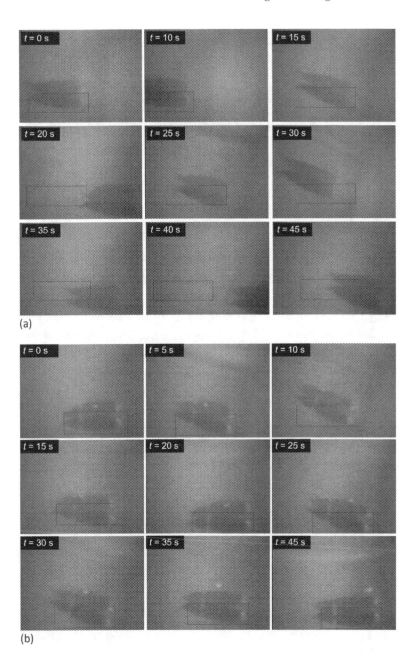

FIGURE 8.6 Two snapshot sequences of the underwater image taken by robotic fish. (a) The case without active vision tracking system. (b) The case with active vision tracking system. Note that the bounding box indicates the detection result of the target.

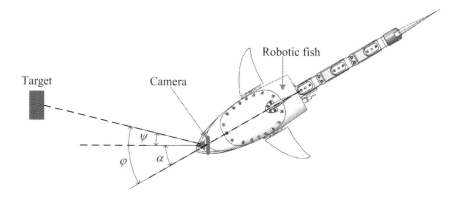

FIGURE 8.7   Position relationship between the target object and the robotic fish.

between the target object and the fish body can then be inferred from the current state of the motor and the acquired image. Here, α denotes the angle between the camera's optical axis and the long axis of the fish body (i.e., longitudinal axis), which is measured through the voltage of potentiometer equipped on the motor shaft of camera stabilizer; ψ indicates the angle between the camera's optical axis and the connection of the camera with the target object, which is calculated from the image based on KCF, the camera model, and its internal parameter; and φ stands for the angle between the long axis of the fish body and the connection of the camera with the target object. As can be easily identified, the following relationship holds: $\varphi = \alpha + \psi$. Thus, an intuitive idea about target tracking control is to make φ as small as possible.

Algorithm 1 DDPG

Randomly initialize the weights θ^Q and θ^μ of Critic network $Q(s, a \mid \theta^Q)$ and Actor network $\mu(s \mid \theta^\mu)$, and then initialize weights $\theta^{Q'}$ and $\theta^{\mu'}$ of the two target networks with θ^Q and θ^μ: $\theta^{Q'} \leftarrow \theta^Q$, $\theta^{\mu'} \leftarrow \theta^\mu$.
Initialize the experience replay pool R
for episode $= 1$ to M **do**
Initialize a random process $\mathcal{N}$ for searching the action space.
Initialize the system and get the initial state s_1
  for $t = 1$ to T **do**
    Choose an action $a_t = \mu(s_t \mid \theta^\mu) + \mathcal{N}_t$ with noise base on the current training strategy.

Execute action a_t to get reward r_t and new state s_{t+1}.

Put this state transition data (s_t, a_t, r_t, s_{t+1}) into the experience playback pool R.

Take a batch of records (s_i, a_i, r_i, s_{i+1}) of size N from the experience playback pool R.

Set $y_i = r_i + \gamma Q'(s_{i+1}, \mu'(s_{i+1} \mid \theta^{\mu'}) \mid \theta^{Q'})$.

Update the Critic network according to the loss function:

$$L = \frac{1}{N} \sum_i (y_i - Q(s_i, a_i \mid \theta^Q))^2$$

Update Actor network according to deterministic policy gradient formula:

$$\nabla_{\theta^\mu} J \approx \frac{1}{N} \sum_i \nabla_a Q(s, a \mid \theta^Q)\Big|_{s=s_i, a=\mu(s_i)} \nabla_{\theta^\mu} \mu(s \mid \theta^\mu) \mid s_i$$

Update network parameters:

$$\theta^{Q'} \leftarrow \delta^Q \theta^Q + (1 - \delta^Q)\theta^{Q'}$$

$$\theta^{\mu'} \leftarrow \delta^\mu \theta^\mu + (1 - \delta^\mu)\theta^{\mu'}$$

end for
end for

Let the yaw angle of the robotic fish be γ. According to the previously presented CPG-based locomotion control, γ can be changed by adjusting the direction-related offset variable β. If $\Delta\gamma = -\Delta\varphi$, the robotic fish will direct toward the target object. In principle, the robotic fish will inevitably produce a displacement affecting φ. In fact, the effect of the displacement can be ignored since the robotic fish is often far from the target object. Hence, we will continuously change yaw angle γ by adjusting β so as to make the robotic fish swim toward the target.

As mentioned previously, RL has the ability to accomplish adaptive controllers without access to accurate dynamical models, making it suitable for underwater robot control. Meanwhile, self-evolution of RL in uncertain environments endows the robot with strong environmental adaptability. Hence, a deep RL-based approach (i.e., DDPG) is adopted to achieve the desired target tracking task [20]. Instead of an accurate dynamic model for fishlike swimming, a general dynamic model for underwater vehicles will be utilized in simulated training. Because we only concern about the yaw angle in this work, the differential equation for the yaw angle can be obtained as follows:

$$a\ddot{\gamma} + b\dot{\gamma} = \tau \tag{8.4}$$

where τ is the vector of thrust forces and moments from the fishlike swimming in the body-fixed frame, which can be simplified as a linear relationship with the exerted offset β. Therefore, we can get

$$\beta = p\ddot{\gamma} + q\dot{\gamma} \tag{8.5}$$

where p and q are the linearized parameters.

Thus, the transfer function from the exerted offset β to the yaw angle γ becomes

$$\frac{\Gamma(s)}{B(s)} = \frac{1}{ps^2 + qs} \tag{8.6}$$

Apparently, it is a second-order system, which can be rewritten as a state equation taking on the following form

$$\begin{cases} \dfrac{d\gamma}{dt} = \dot{\gamma} \\[2mm] \dfrac{d\dot{\gamma}}{dt} = \dfrac{\beta}{p} - \dfrac{q}{p}\dot{\gamma} \end{cases} \tag{8.7}$$

That is, the system state can be fully described by a bivector $(\gamma, \dot{\gamma})$, and the state of the system at the next moment is determined only by γ, $\dot{\gamma}$, and β. Hence, choosing $(\gamma, \dot{\gamma})$ as the state vector can ensure that the target tracking control problem meets the conditions of Markov decision processes (MDP).

Within the DDPG framework [20], the output value needs to be normalized. However, the actually exerted offset should not be too large; otherwise there will be a nonlinear relationship between β and γ. In order to limit the DDPG output size, we set

$$\beta = k_\beta \beta' \tag{8.8}$$

where k_β denotes the discount factor, whose range is $(0,1)$. (β') is selected as the action vector and then the range of β becomes $(0, k_\beta)$.

Further, a well-designed, heuristic reward function (r_γ) will be designed. The simplest idea is to use the current tracking error as a reward function. That is, the larger the tracking error, the lower the return value. Specifically, the tracking error–based reward function takes the following form:

$$r_\gamma = \begin{cases} \dfrac{2-|\gamma|}{2} & |\gamma| \le 2 \\ 0 & \text{others} \end{cases}. \tag{8.9}$$

Direct use of equation (8.9) as a reward function makes the system prone to overshoot and oscillation. In order to solve the oscillating problem and make the control system try not to overshoot, angular velocity is also introduced to the reward function. As with the angular velocity, the system does not want to make the angular velocity smaller in any case. On the contrary, when the tracking error is large, a larger angular velocity will shorten the rise time. When the tracking error gets small, a smaller angular velocity can make the overshoot smaller and the transition smoother, as well as suppress the occurrence of system output oscillation. For this purpose, the angular velocity–based reward function is designed as a piecewise function:

$$r_{\dot\gamma} = \begin{cases} 1-|\dot\gamma| & |\dot\gamma| \le 1, |\gamma| \le 0.1 \\ 0 & \text{others} \end{cases}. \tag{8.10}$$

Further, a weighted combination of the tracking error–based and angular velocity–based reward functions yields

$$r = \frac{w_\gamma r_\gamma + w_{\dot\gamma} r_{\dot\gamma}}{w_\gamma + w_{\dot\gamma}} \tag{8.11}$$

where w_γ and $w_{\dot\gamma}$ represent the weights of the tracking error–based and angular velocity–based reward functions, respectively. It should be remarked that using weights in the formation of the reward function facilitates the balance of different metrics. In this work, two weights are set as $w_\gamma = 2$ and $w_{\dot\gamma} = 1$, respectively. Note also that equations (8.9)–(8.11) deliberately maintain the range of the reward function as [0,1], because the normalized reward function can maximally guarantee the convergence of the DDPG algorithm without the need to modify hyperparameters and network structure [20]. In particular, the termination condition of the control system is set as $|\gamma| > 2$. Once the termination condition is triggered, the return function will be set to a fixed value $r = -10$.

In brief, the overall structure of the proposed underwater target tracking control system is illustrated in Figure 8.8. In this system, the robotic fish is governed by the CPG-based controller to achieve fishlike swimming. The underwater image is sent to a host computer onshore and processed by KCF algorithm to detect the position of target in the image plane. The angle ψ is further obtained by means of pinhole camera model, which is inputted to a camera stabilizer to keep the camera toward the target. After that, the angle α of the potentiometer and the angle ψ are combined to calculate the expected yaw angle and its difference of the robotic fish. Then, the yaw angle serves as the input of the DDPG to make sure how the robotic fish turns, which is decided by the offset parameter of CPG. Thereafter, the robotic fish can turn to track the underwater target. Specifically, the relative position (γ) of the target to the robotic fish and its derivative $(\dot{\gamma})$ constitute the state vector of the DDPG in the tracking task, which are the inputs of Actor network. As for the yaw motion of the

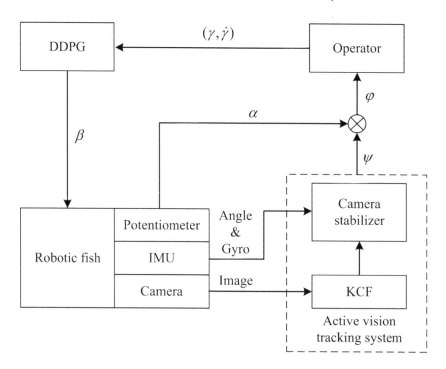

FIGURE 8.8 The overall structure of the target tracking control system.

robotic fish, it can be controlled by the CPG model bias rate (β). Since the DDPG normalizes the behavior space, we introduce the transition variable (β') as the action of the DDPG, which is the output of Action network. Hence, CPG control–related parameter (β) can be connected to the DDPG output (β') with its coefficient $\left(k_\beta\right)$.

8.4.2 Performance Analysis of DDPG-Based Control System

In order to verify the effectiveness of the DDPG algorithm, as shown in Figure 8.9, a simulated learning system is built and evaluated. Specifically, equation (8.6) is used as the transfer function of the controlled system, equation (8.11) as the reward function, $(\gamma,\dot{\gamma})$ as the state vector, and (β') as the action vector, where $p=1$, $q=1.25$, $k_\beta=0.5$. In the simulations, the network structure and hyperparameters of the DDPG algorithm are consistent with those in [20]. The simulation control period is 0.25 s, and the maximum number of simulation steps is 200 steps. Figure 8.10 displays the variation of the total reward over the number of steps in a training episode. As can be observed, the total reward has barely changed before 10,000 steps and has been hovering around zero. This is because a warm-up procedure of the Q network is executed, exactly corresponding to 10,000 steps. It should be remarked that the so-called warm-up is the initial training of the Critic network rather than the Actor network for a period of time. After the Critic network has been trained to a certain degree, the Actor network begins training. The warm-up of the DDPG can avoid random and even erroneous training of the Actor network until the Critic network achieves a good estimate of system characteristics, thereby speeding up the convergence of the algorithm. As can be seen from Figure 8.10, the algorithm converges to the optimal solution at about 20,000 steps. That is, the system has been running 100 times to find the optimal solution,

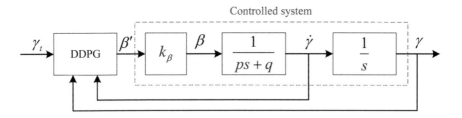

FIGURE 8.9 Block diagram of the DDPG-based learning system.

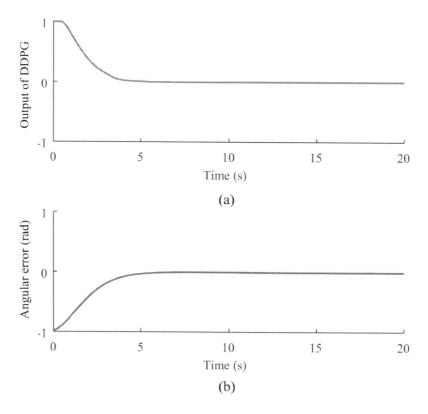

FIGURE 8.10   Plot of the total reward over the number of steps in a training episode.

revealing a high convergence speed. Figure 8.11 further plots the step response of the control system under the DDPG controller. Apparently, the system adapts quickly and does not overshoot. The transient period for this system is approximately 5 s, satisfying the design expectations of the controller.

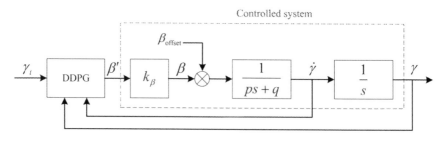

FIGURE 8.11   Step response of the DDPG-based learning system in simulation. (a) Output of DDPG. (b) Angular error.

Next, the stability analysis of the DDPG-based learning system is performed since the deep Q learning is prone to be unstable in certain circumstances. As mentioned previously, instead of an accurate dynamic model of fishlike swimming, the utilization of a general dynamic model for underwater vehicles may lead to instability of the learning system. In order to analyze the stability of the simulation control system, the Bode plot of its open-loop transfer function can be employed to obtain its gain margin and phase margin so as to judge its stability qualitatively. Simulation analysis of the learning system is performed in Matlab environment. According to simulation results, the gain margin is 51.5 dB (at 13.4 rad/s) and the phase margin is 85.3° (at 0.343 rad/s), indicating that this system can run stably in the actual control system.

Besides the control system stability analysis, the adaptability of the learning system is examined especially when the controlled system changes. A typical case is a static bias existing in the moving joints of the robotic fish tail, which will severely affect the straight swimming performance. It is assumed that the oscillating joint has a static bias of 0.2, i.e., $\beta_{offset} = 0.2$. For an actual robotic fish, this static bias is not higher than 0.2, usually less than 0.1. The main reason for choosing a higher static bias is to make the control effect more pronounced. The control system influenced by the static bias is illustrated in Figure 8.12, whose step response is plotted in Figure 8.13. Note that the size of the used replay buffer in running DDPG algorithm is 50,000, representing a tradeoff between total reward and convergence speed. As can be observed from Figure 8.13, in spite of a

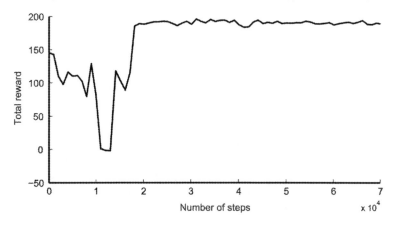

FIGURE 8.12   Block diagram of the DDPG-based learning system influenced by static bias.

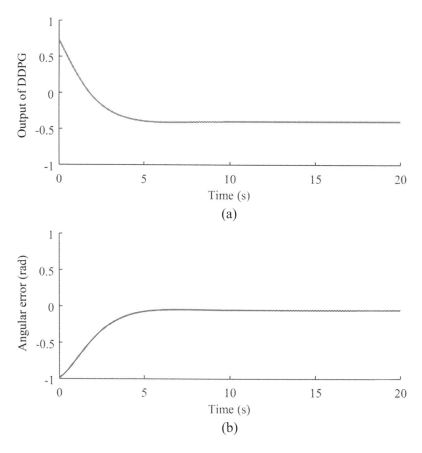

FIGURE 8.13 Step response of the DDPG-based learning system influenced by static bias. (a) Output of DDPG. (b) Angular error.

steady-state error in the system, the learning system can be fully adapted to the imposed static bias.

In addition, the time-delay issue of the learning system is explored. During the target tracking, the actual control system will be influenced by different hysteresis elements involving filters, image acquisition, CPG-based control, etc. These hysteresis elements can be regarded as first-order inertia elements. The sum of the time constants of these inertial elements can reach 1.5–2 s. The presence of hysteresis elements will greatly affect the stability of the system and may even make the controller trained in the simulation unusable in the real environment. For this reason, some theoretical analysis should be made to evaluate

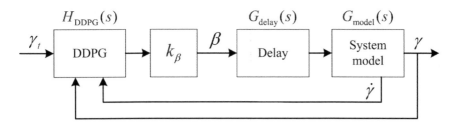

FIGURE 8.14   Block diagram of the DDPG-based learning system influenced by time delays.

the adverse effects of hysteresis elements. For simplicity, three typical hysteresis elements from filters, image acquisition, and CPG-based control are regarded as an equivalent hysteresis element, whose open-loop transfer function is described by

$$G_{\text{delay}} = e^{-T_{\text{delay}} s} \approx \frac{1}{T_{\text{delay}} s + 1} \tag{8.12}$$

where T_{delay} denotes the lag time of the system.

Figure 8.14 illustrates the block diagram of the simplified DDPG-based learning system with time delays. Let $T_{\text{delay}} = 2$. The corresponding step response of this simplified learning system is plotted in Figure 8.15. As can be identified, as opposed to the normal state illustrated in Figure 8.11, the adjustment time of the learning system with time delays gets longer, around 13 s. Notably, the learning system still meets the operational stability requirement. From the viewpoint of control theory, after incorporating an equivalent hysteresis element, the learning system becomes a third-order one. The state vector $(\gamma, \dot{\gamma})$ hardly describes the system state completely. Therefore, the initial MDP condition is no longer guaranteed to hold. In fact, the controlled system with an equivalent hysteresis element may convert to a partially observable Markov decision process (POMDP). Fortunately, under certain circumstances, RL can be used to solve POMDPs so as to eliminate the limitation of a 2D state-space structure [29].

8.5 EXPERIMENTS AND RESULTS

In order to verify the effectiveness of the proposed RL-based target tracking control for the vision-guided robotic fish, aquatic experiments were carried out in an indoor swimming pool, whose dimension is 5 m×4 m×1.5 m. In particular, a specialized motion measurement system based

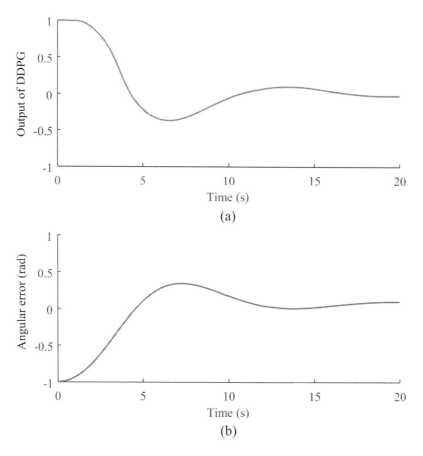

FIGURE 8.15 Step response of the DDPG-based learning system influenced by time delays. (a) Output of DDPG. (b) Angular error.

on a global vision system was adopted [30]. The global camera was suspended from the ceiling with a height of approximately 1.9 m from the water surface. Note that the global camera is only used to record the whole experiment but not provide any information to the robotic fish for target tracking. A dedicated tracking program was running on the host computer, which was responsible for continuous detection and tracking of the colored target object. Accordingly, the position and speed information about the target could be available in a real-time manner.

8.5.1 Static and Dynamic Tracking Experiments

The first experiment concerned the step response of the physical system. As demonstrated previously, the DDPG-based learning system could

maintain the stability of the system as a whole, even if various disturbances that may exist in the physical system were imposed on the controller. It is feasible to use the characteristic parameters of the learning system
trained in the simulation environment as the pre-training parameters
of the RL-based tracking controller in the physical environment. Thus,
the learning system could be continuously evolved toward seeking optimal solutions for target tracking problems, while maintaining the overall control system stable. The step response of the real learning system
in aquatic environments is shown in Figure 8.16, where the robotic fish
was required to detect the static target object represented by a dolphin
robot [31]. Snapshots of the step response test are given in Figure 8.17. It
should be remarked that although the global visual motion measurement

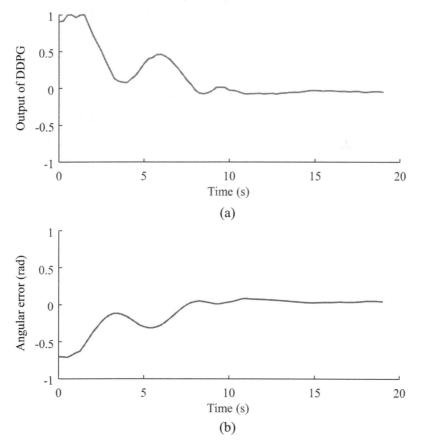

FIGURE 8.16   Step response of the DDPG-based learning system in real aquatic
environments. (a) Output of DDPG. (b) Angular error.

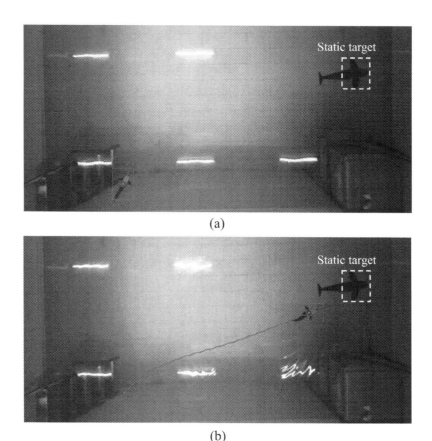

(a)

(b)

FIGURE 8.17   Snapshots of the step response test. (a) The initial position and direction of the robotic fish. (b) The plotted trajectory of the robotic fish in the whole step response process.

system was used in the experiments, the global visual information was not employed in motion planning and control of the robotic fish. A careful inspection reveals that the adjustment time of this step response experiment is about 10 s, lying between the simulated case without disturbance being approximately 5 s (see Figure 8.11) and the simulated case considering the effect of time delays being approximately 13 s (see Figure 8.15). At the same time, the output of the control system is not oscillating, whose static error is less than 0.1; the real trajectory of the robotic fish is fairly smooth, agreeing well with the simulated results.

In order to further verify the necessity of RL, we conducted a comparative experiment between DDPG and conventional PID controller. One

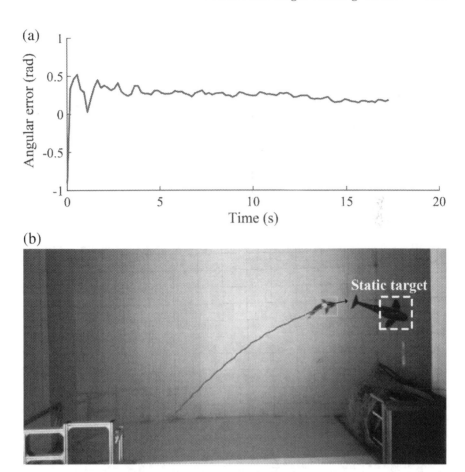

FIGURE 8.18 The step response test under control of the PID method. (a) Plot of the angular error. (b) The plotted trajectory of the robotic fish in the whole process.

of the experimental results on PID controller is shown in Figure 8.18a. Comparing Figure 8.16b and Figure 8.18a, we can observe that the transient response of PID controller is much faster than DDPG, while the overshoot and steady-state error is much worse. PID controller only takes about 1 s to reach peak value, but its overshoot is over 0.5 rad, and it always has a steady-state error at 0.2 rad. Increasing the weight of the integral part can slightly decrease the steady-state error, but in exchange the tracking system becomes instable, and the success rate of tracking decreases sharply. As for DDPG, although the transient response is worse than PID controller, the tracking process is smoother and more stable with no steady-state error. In addition, a careful inspection of Figure 8.18b reveals that the tracking trajectory of DDPG is obviously shorter than PID controller because DDPG with no steady-state error adjusts the robotic fish's straight swimming toward the target whereas PID controller governs the swimming of robotic fish in a curve. In brief, DDPG performs better than PID controller. Additionally, its performance can further be improved through interaction with environments.

In another experiment, the robotic fish was required to track a dynamic target object. As an illustrative case, Figure 8.19 offers a snapshot sequence of successful tracking of a moving dolphin robot. The overall test lasted about 40 s. As can be observed, the robotic fish could detect and follow

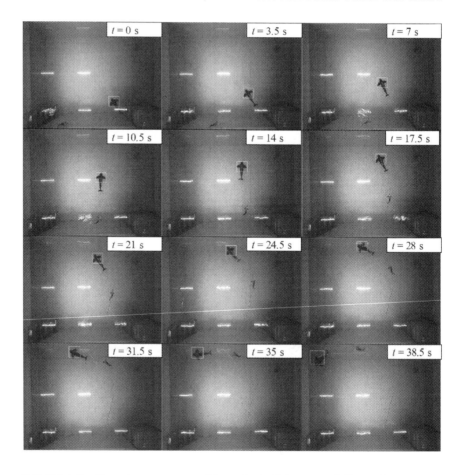

FIGURE 8.19 Snapshot sequence of the robotic fish successfully tracking a moving dolphin robot. Note that the snapshot interval time is set as 3.5 s here.

the free-swimming robotic dolphin. Notably, according to the moving object trajectories, when the movement of the robotic dolphin altered, the robotic fish could adjust its direction in due course. Figure 8.20 further gives curves of target angle and control output as well as their angular error for the dynamic tracking experiment. Note that a duration of 20 s featuring stable target tracking is chosen. As can be observed from Figure 8.20a, although the system response lagged a few behind the target angle, it could still follow the change of target angle successfully. A careful inspection of Figure 8.20b reveals that the angular error fluctuated around 0.5 rad. Considering the large lag in image transmission, this kind of tracking performance is acceptable in the context of underwater task execution. In this sense, the success of dynamic tracking experiment reflected the

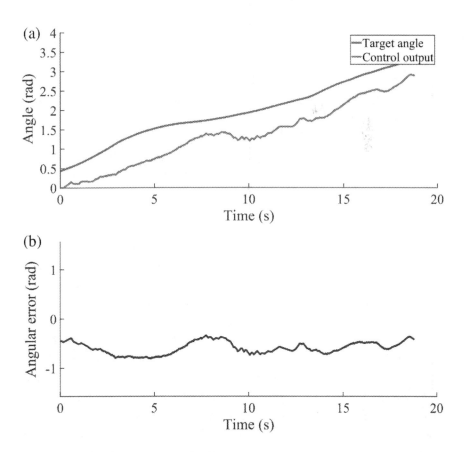

FIGURE 8.20 Performance analysis of the DDPG-based learning system in the dynamic case. (a) Plot of target angle and control output. (b) Plot of angular error.

effectiveness and adaptability of the DDPG-based learning system in handling dynamic tracking problem.

8.5.2 Discussion

Robotic vision systems are widely utilized for air, land, and sea-oriented target tracking tasks [32–34]. Owing to the special properties of aquatic environment and robot motion, underwater target tracking is faced with a huge challenge in creating flexible and robust tracking algorithms. In particular, tracking algorithms under dynamic situations should be adaptive with changing scenes, robust to noisy images, and capable of real-time implementation. The proposal of the novel camera stabilizing system and the DDPG-based learning system allows the robotic fish to track image-based target tracking flexibly. Here the target tracking problem can be converted into the states of image stabilization, deep learning, and continuous control. In contrast to the aforementioned research on tracking control of free-swimming robotic fish [13–16], two improvements are highlighted in this work. On the one hand, the creation of the camera stabilizer aided by the feedback-feedforward controller permits stable and clear underwater image acquisition, paving the way for image-based target tracking solidly. On the other hand, the DDPG-based learning control in tracking parameter tuning not only reduces the dependency on accurate dynamic models of fish swimming to a large extent but also ensures the speediness of the proposed target tracking control. Unlike the MPF-type boxfish-like robot serving as the aquatic verification platform [13], the BCF-type robotic fish is intrinsically instable in yaw motions and image-based control. With the assistance of the active visual tracking system and corresponding tracking control algorithms, the robotic fish is endowed with the ability to stably track moving target in the 2D plane. Accidently, to the best of our knowledge, wide-range moving target tracking is achieved on the BCF-type untethered robotic fish for the first time.

Despite successfully implementing vision-based target tracking with self-propelled robotic fish, the proposed control methods have some limitations. First, the image stabilization scheme based on one-DoF camera stabilizer is compact, yet unable to suppress slight roll jitter. As a prerequisite of a miniaturized integrated propulsion platform with a fishlike shape, designing two-DoF or even three-DoF camera stabilizers as much as possible will make the image stabilization effect better. Second, dramatic changes in target trajectory go along with a large hysteresis of the tracking system. The primary cause is that image acquisition and processing bring a pure hysteresis element with a large time constant for the overall active vision tracking system. In practice, this pure hysteresis element is likely to cause the active vision tracking system to be extremely unstable. Thus, it needs to determine the time constant of this pure hysteresis and further to develop corresponding compensation algorithms for better tracking performance. At last, the convergence time of the employed DDPG-based learning system is a bit long, possibly leading to the instability of the overall control system during the transition process. To avoid this dilemma, one possible alternative is to explore enhanced learning control algorithms that are well suited to bioinspired swimming and real-time implementation.

8.6 CONCLUSIONS AND FUTURE WORK

In this chapter, we have offered a free-swimming robotic fish-based solution to underwater target tracking using onboard vision. Specifically, unlike passive installation of an embedded camera in the fish head, a novel active visual tracking system with the inclusion of a camera stabilizer is firstly proposed. Then, a cascade control framework consistent with the physical configuration of the active visual tracking system is built. On the basis of vision-based tracking control, a DDPG-based target tracking controller is developed, endowing the robotic fish with the capability of smooth and stable tracking of the target object. Finally, the experiments on static and dynamic target tracking illustrate the reasonable performance of the proposed mechatronic design and control methods. As is clearly shown by both simulated and real experiments, vision-guided robotic fish aided by RL is competent to the task of dynamic target tracking. Remarkably, such results may offer insight into parameter tuning and control update of onboard vision-guided tracking tasks in changing environments. On the other hand, the DDPG in work is applied to cope with the robotic control problem but not image processing. Another alternative is to directly input image into DRL algorithm and make the tracking system as an end-to-end architecture. Thus, the DRL can cope with complex target tracking missions through interaction with the environments, such as in conditions with low or no light in the ocean or a lake.

Our future work will concentrate on 3D tracking control of robotic fish under dynamic aquatic environments. Besides the improvement of the camera stabilizer, more attention will be paid to the exploration of other learning control algorithms that are more suitable for multiple fast-moving and agile robotic fishes [35]. We will also further explore the applications of RL in robotic fish control, such as taking unprocessed underwater images as input directly. It is anticipated that the bioinspired robotic fishes can be applied in various tasks, including observation, exploration, surveillance, and even operation in virtue of manipulators.

REFERENCES

1. A. J. Murphy and M. Haroutunian, "Using bio-inspiration to improve capabilities of underwater vehicles," in *Proceedings of the 17th International Symposium on Unmanned Untethered Submersible Technology*, Portsmouth, New Hampshire, USA, Aug. 2011, pp. 20–31.
2. L. Wen, T. Wang, G. Wu, and J. Liang, "Quantitative thrust efficiency of a self-propulsive robotic fish: Experimental method and hydrodynamic investigation," *IEEE/ASME Trans. Mechatron.*, vol. 18, no. 3, pp. 1027–1038, 2013.
3. Y. Wang, X. Yang, Y. Chen, D. K. Wainwright, C. P. Kenaley, Z. Gong, Z. Liu, H. Liu, J. Guan, T. Wang, J. C. Weaver, R. J. Wood, and L. Wen, "A biorobotic adhesive disc for underwater hitchhiking inspired by the remora suckerfish," *Sci. Rob.*, vol. 2, no. 10, p. eaan8072, 2017.
4. W. Wang, D. Gu, and G. Xie, "Autonomous optimization of swimming gait in a fish robot with multiple onboard sensors," *IEEE Trans. Syst. Man Cybern. Syst.*, vol. 49, no. 5, pp. 891–903, 2019.
5. M. Sfakiotakis, D. M. Lane, and J. B. C. Davies, "Review of fish swimming modes for aquatic locomotion," *IEEE J. Oceanic Eng.*, vol. 24, no. 2, pp. 237–252, 1999.
6. F. E. Fish, "Advantages of natural propulsive systems," *Mar. Technol. Soc. J.*, vol. 47, no. 5, pp. 37–44, 2013.
7. A. Raj and A. Thakur," Fish-inspired robots: Design, sensing, actuation, and autonomy–A review of research," *Bioinspiration Biomimetics*, vol. 11, no. 3, p. 031001, 2016.
8. J. Yu, M. Wang, H. Dong, Y. Zhang, and Z. Wu, "Motion control and motion coordination of bionic robotic fish: A review," *J. Bionic Eng.*, vol. 15, no. 4, pp. 579–598, 2018.
9. X. Chen, J. Yu, Z. Wu, Y. Meng, and S. Kong, "Towards a maneuverable miniature robotic fish equipped with a novel magnetic actuator system," *IEEE Trans. Syst. Man Cybern. Syst.*, vol. 50, no. 7, pp. 2327–2337, Jul. 2020. DOI: 10.1109/TSMC.2018.2812903.
10. J. Yu, Z. Su, M. Wang, M. Tan, and J. Zhang, "Control of yaw and pitch maneuvers of a multilink dolphin robot," *IEEE Trans. Rob.*, vol. 28, no. 2, pp. 318–329, 2012.

11. S. B. Behbahani and X. Tan, "Design and modeling of flexible passive rowing joints for robotic fish pectoral fins," *IEEE Trans. Rob.*, vol. 32, no. 5, pp. 1119–1132, 2016.

12. J. Yu, C. Wang, and G. Xie, "Coordination of multiple robotic fish with applications to underwater robot competition," *IEEE Trans. Ind. Electron.*, vol. 63, no. 2, pp. 1280–1288, 2016.

13. Y. Hu, W. Zhao, and L. Wang, "Vision-based target tracking and collision avoidance for two autonomous robotic fish," *IEEE Trans. Ind. Electron.*, vol. 56, no. 5, pp. 1401–1410, 2009.

14. Y. Takada, K. Koyama, and T. Usami, "Position estimation of small robotic fish based on camera information and gyro sensors," *Robotics*, vol. 3, no. 2, pp. 149–162, 2014.

15. J. Yu, F. Sun, D. Xu, and M. Tan, "Embedded vision guided 3-D tracking control for robotic fish," *IEEE Trans. Ind. Electron.*, vol. 63, no. 1, pp. 355–363, 2016.

16. F. Sun, J. Yu, P. Zhao, and D. Xu, "Tracking control for a biomimetic robotic fish guided by active vision," *Int. J. Rob. Autom.*, vol. 31, no. 2, pp. 137–145, 2016.

17. D. D. Doyle, A. L. Jennings, and J. T. Black, "Optical flow background estimation for real-time pan/tilt camera object tracking," *Measurement*, vol. 48, pp. 195–207, 2014.

18. J. Kober, J. A. Bagnell, and J. Peters, "Reinforcement learning in robotics: A survey," *Int. J. Rob. Res.*, vol. 32, no. 11, pp. 1238–1274, 2013.

19. D. Michie and R. A. Chambers, "Boxes: An experiment in adaptive control," in *Machine Intelligence 2*, E. Dale and D. Michie, Eds. Edinburgh: Oliver and Boyd, 1968, pp. 137–152.

20. T. P. Lillicrap, J. J. Hunt, A. Pritzel, N. Heess, T. Erez, Y. Tassa, and D. Silver, "Continuous control with deep reinforcement learning," *Computer Sci.*, vol. 8, no. 6, A187, 2015.

21. Z. Shen, M. Tan, Z. Cao, S. Wang, and Z. Hou, "Obstacle avoidance learning for biomimetic robot fish," *Appl. Artif. Intell.*, pp. 719–724, 2006. https://www.worldscientific.com/doi/10.1142/9789812774118_0100

22. L. Lin, H. Xie, D. Zhang, and L. Shen, "Supervised neural Q-learning based motion control for bionic underwater robots," *J. Bionic Eng.*, vol. 7, Suppl., pp. S177–S184, 2010.

23. X. Yang, Z. Wu, and J. Yu, "Design and implementation of a robotic shark with a novel embedded vision system," in *Proceedings of the IEEE International Conference on Robotics and Biomimetics*, Qingdao, China, Dec. 2016, pp. 841–846.

24. J. Yu, S. Chen, Z. Wu, and W. Wang, "On a miniature free-swimming robotic fish with multiple sensors," *Int. J. Adv. Rob. Syst.*, vol. 13, no. 2, 62, 2016.

25. J. Yu, M. Tan, J. Chen, and J. Zhang, "A survey on CPG-inspired control models and system implementation," *IEEE Trans. Neural Networks Learn. Syst.*, vol. 25, no. 3, pp. 441–456, 2014.

26. A. J. Ijspeert, A. Crespi, D. Ryczko, and J. M. Cabelguen, "From swimming to walking with a salamander robot driven by a spinal cord model," *Science*, vol. 315, no. 5817, pp. 1416–1420, 2007.

27. J. F. Henriques, R. Caseiro, P. Martins, and J. Batista, "High-speed tracking with kernelized correlation filters," *IEEE Trans. Pattern Anal. Mach. Intell.*, vol. 37, no. 3, pp. 583–596, 2015.

28. X. Yang, Z. Wu, and J. Yu, "A novel active tracking system for robotic fish based on cascade control structure," in *Proceedings of the IEEE International Conference on Robotics and Biomimetics*, Qingdao, China, Dec. 2016, pp. 749–754.

29. P. Zhu, X. Li, P. Poupart, and G. Miao, "On improving deep reinforcement learning for POMDPs," https://arxiv.org/abs/1704.07978, 2018.

30. J. Yuan, J. Yu, Z. Wu, and M. Tan, "Precise planar motion measurement of a swimming multi-joint robotic fish," *Sci. China Inf. Sci.*, vol. 59, no. 9, p. 092208, 2016.

31. Z. Wu, J. Liu, J. Yu, and H. Fang, "Development of a novel robotic dolphin and its application to water quality monitoring," *IEEE/ASME Trans. Mechatron.*, vol. 22, no. 5, pp. 2130–2140, 2017.

32. H. Xu and Y.-P. Shen, "Target tracking control of mobile robot in diversified manoeuvre modes with a low cost embedded vision system," *Ind. Rob.*, vol. 40 no. 3, pp. 275–287, 2013.

33. E. H. C. Harik, F. Guerin, F. Guinand, J.-F. Brethe, H. Pelvillain, and J. Y. Parede, "Fuzzy logic controller for predictive vision-based target tracking with an unmanned aerial vehicle," *Adv. Rob.*, vo. 31, no. 7, pp. 368–381, 2017.

34. A. Jebelli and M. C. E. Yagoub, "Efficient robot vision system for underwater object tracking," in *Proceedings of the 2nd International Conference on Control Science and Systems Engineering*, Singapore, Jul. 2016, pp. 242–247.

35. Z. Ji and H. Yu, "A new perspective to graphical characterization of multi-agent controllability," *IEEE Trans. Cybern.*, vol. 47, no. 6, pp. 1471–1483, 2017.

CHAPTER 9

Summary and Outlook

T HIS BOOK FOCUSING ON the underwater visual perception and control technology involves four primary parts, i.e., underwater image restoration, calibration methodology for underwater accurate stereo measurement, underwater object detection by artificial intelligent tools, and related applications which combine the visual methods with advanced control methods on the underwater robot platforms.

With respect to the highly efficient underwater image restoration, an adaptive real-time underwater visual restoration with adversarial critical learning is proposed. In particular, we develop a multi-branch discriminator including an adversarial branch and a critic branch for the purpose of simultaneously preserving image content and removing underwater noise. In addition to adversarial learning, a novel dark channel prior loss is designed to promote the generator to produce a realistic vision. More specifically, an underwater index is investigated to describe underwater properties, and a loss function based on the underwater index is designed to train the critic branch for underwater noise suppression. Through extensive comparisons on visual quality and feature restoration, the superiority of the proposed approach is confirmed. Consequently, GAN-RS can adaptively improve underwater visual quality in real time and yield an overall superior restoration performance.

Looking at the calibration method for the underwater stereo measurement system, an NSGA-II-based calibration algorithm is provided in this book in order to improve the underwater measurement accuracy. In this situation, a refractive camera model and an akin triangulation are proposed to establish the nonlinear relationship with housing parameters between the object and its corresponding image plain points. A novel usage of checkerboard based on the relative position relationship of corners is employed to set three optimal goals, i.e., the distance difference, the vertical direction difference, and the parallel direction difference. The process of calibration is regarded as a multi-objective optimization and solved by NSGA-II. Finally, experimental results demonstrate the validity and effectiveness of the proposed calibration algorithm.

Due to that object detection is a crucial technique for the environment understanding of underwater robots, this book emphasizes the technologies of object detection. First, we propose a joint anchor-feature refinement for real-time accurate object detections. As a dual refinement mechanism, a novel anchor-offset detection is designed, which includes an anchor refinement, a feature location refinement, and a deformable detection head. This new detection mode is able to simultaneously perform two-step regression and capture accurate object features. Based on the anchor-offset detection, a dual refinement network (DRNet) is developed for high-performance static detection, where a multi-deformable head is further designed to leverage contextual information for describing objects. As for temporal detection in videos, temporal refinement networks (TRNet) and temporal dual refinement networks (TDRNet) are developed by propagating the refinement information across time. Second, the rethinking temporal object detection method is implemented for robotic applications. In this book, non-reference assessments are proposed for continuity and stability based on object tracklets. These temporal evaluations can serve as supplements to static AP. Further, we develop an online tracklet refinement for improving detectors' temporal performance through short tracklet suppression, fragment filling, and temporal location fusion. In addition, we propose a small-overlap suppression to extend VID methods to single object tracking (SOT) task so that a flexible SOT-by-detection framework is then formed.

Notably, this book creatively studies the relationship between image restoration and detection accuracy. We generally investigate the relation of quality-diverse data domain to detection performance. In the meantime, we unveil how visual restoration contributes to object detection in real-world underwater scenes. According to our analysis, five key discoveries are reported: (1) Domain quality has an ignorable effect on within-domain convolutional representation and detection accuracy; (2) low-quality domain leads to higher generalization ability in cross-domain detection; (3) low-quality domain can hardly be well learned in a domain-mixed learning process; (4) degrading recall efficiency, restoration cannot improve within-domain detection accuracy; (5) visual restoration is beneficial to detection in the wild by reducing the domain shift between training data and real-world scenes.

Additionally, there are two representative underwater robotic practices, which integrate the visual perception technologies and advanced control methods, introduced in this book. With respect to the surface environment, a robot system for intelligent water surface cleaner named IWSCR is developed to collect floating plastic garbage. It is able to accomplish three major tasks autonomously, i.e., cruise and detection, tracking and steering, and grasping and collection. First of all, the YOLOv3 network, which is widely applied in the high speed and the accuracy object detection field, is trained on the proposed floating garbage dataset to realize accurate and real-time garbage detection. Next, to improve the ability of resisting disturbances, a control law based on sliding mode controller is proposed for vision-based steering. Furthermore, inspired by the stability of floating bottles in fluid, a feasible grasping strategy is utilized for IWSCR. Focusing on the underwater environment, we present a novel robotic fish platform with a camera stabilizing system and achieve real-time 2D target tracking assisted by reinforcement learning (RL) in continuous environments. More specifically, we firstly develop an active visual tracking system based on cascade control structure to obtain the relative orientation between the robotic fish and the underwater target. Then, we propose a target tracking controller dealing with continuous state and action spaces based on deep reinforcement learning (DRL). The controller takes the position of the target object as input and yields the motion parameters of the bioinspired central pattern generator governed robotic fish. The robustness and adaptability of the proposed controller as well as the influence of time-delays on the control system are explored via simulated experiments under different scenarios.

It is an exact tendency that a higher degree of automation and intelligence will be implemented on the underwater robotic systems in the future. Visual perception methods will be investigated further toward a lightweight and adaptable direction, by which underwater robots are capable of exploring the uncharted ocean tackling the computation limitation and environment uncertainties. The increasing number of artificial intelligent technologies will be applied in the underwater robotic fields in order to accomplish more feasible operation tasks.

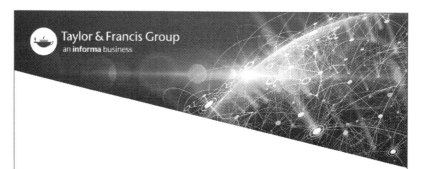

Printed in the United States
by Baker & Taylor Publisher Services